CONNECTED IN ISOLATION

CONNECTED IN ISOLATION

DIGITAL PRIVILEGE IN UNSETTLED TIMES

ESZTER HARGITTAI

The MIT Press
Cambridge, Massachusetts
London, England

The MIT Press would like to thank the anonymous peer reviewers who provided comments on drafts of this book. The generous work of academic experts is essential for establishing the authority and quality of our publications. We acknowledge with gratitude the contributions of these otherwise uncredited readers.

This book was set in Bembo Book MT Pro by Westchester Publishing Services. Printed and bound in the United States of America.

Library of Congress Cataloging-in-Publication Data

Names: Hargittai, Eszter, 1973– author.
Title: Connected in isolation : digital privilege in unsettled times / Eszter Hargittai.
Description: Cambridge, Massachusetts : The MIT Press, [2022] | Includes bibliographical references and index.
Identifiers: LCCN 2022005156 (print) | LCCN 2022005157 (ebook) |
    ISBN 9780262047371 (paperback) | ISBN 9780262371490 (pdf) |
    ISBN 9780262371506 (epub)
Subjects: LCSH: Digital divide. | Digital media. | Information society—Social aspects. |
    Covid-19 (Disease)—Social aspects.
Classification: LCC HM851 .H3665 2022 (print) | LCC HM851 (ebook) |
    DDC 303.48/33—dc23/eng/20220207
LC record available at https://lccn.loc.gov/2022005156
LC ebook record available at https://lccn.loc.gov/2022005157

10 9 8 7 6 5 4 3 2 1

To my mother, Dr. Magdolna Hargittai, for introducing me to the Internet in 1992, and for so much more.

CONTENTS

PREFACE

Every single person reading this book within several years of its publication will have experienced the lockdowns that form its social context. One of the many unique aspects of conducting research during this time was that lockdown orders threw into disarray the lives of researchers just as they did the lives of the people they were studying. I and my research team did all of the project preparation while we were sheltering in place in our respective homes. Here, I want to recognize these circumstances explicitly as they posed distinctive challenges that are not often part of the research process.

Work under special conditions is certainly not unprecedented and has happened in other instances, such as scholars based in New York City interviewing their fellow New Yorkers immediately after 9/11 or researchers documenting political uprisings while participating in them.[1] Such situations can result in unreproducible observations, which tend to be anathema to the scientific enterprise. Yet under certain conditions, capturing the uniqueness of the moment is itself an important contribution. The limitations under which researchers operate in such unreproducible cases is fortunately not the norm of scholarly work. To recognize the unusual circumstances of research during lockdown, this preface foregrounds the conditions under which my research team conducted this data-collection work. Although it is more common for qualitative scholars, and especially ethnographers, to reflect explicitly on what it is that they bring to a study given their particular social positions, because 2020 challenged everybody, I believe it to be relevant for a survey-based study as well. Fortunately, I wrote some of the reflections in this preface at the time of lockdowns.[2] This is relevant, because as the pandemic drew on, it became easy to forget what our initial impressions and experiences had been like and how unusual the circumstances were at first, even if over time we may had become more accustomed to them.

I collected the data for this book in collaboration with my research team at the University of Zurich. When I use the plural form of the first-person pronoun, it is to recognize the joint efforts of my team members, which were considerable in the data-collection process. Zurich, Switzerland, went into a soft lockdown on March 16, 2020. I myself had been visiting my parents in Budapest, Hungary, the previous week when it became clear that lockdowns were about to take effect. I changed my return ticket to avoid possible complications of getting stuck away from home. On the afternoon of Friday, March 13, our employer, the University of Zurich, let us know that as of Monday, March 16, we would be teaching remotely. This left me the weekend to make the necessary changes to my teaching while rushing back to Zurich on Saturday, March 14, from Budapest.

Everyone on my team—two postdocs, three PhD students, and one master's student—spent the first few days finding our bearings. Most of us were not originally from Switzerland so frequent communication with faraway loved ones was top of mind, as was obsessively following the news. As communication scholars who study Internet uses, this experience was not only personal, it also prompted research questions. How was the pandemic affecting interpersonal communication? How was the public keeping abreast of all the government rules and health recommendations? How were people of different life stages and home circumstances coping? And in particular, what role were digital media, our specific focus of study and area of expertise, playing in all of this?

But research questions were not the only ones swirling in our heads. I wondered, were we in the right frame of mind to launch into a study at all or was it more realistic to focus our efforts on handling this unprecedented situation? Would moving our teaching online, adapting to lockdown, and figuring out life under these new circumstances use up all of our energies? Was it realistic to put so much effort into research when there were so many distractions about simply functioning day to day? Or could we muster efforts for a study about the pandemic?

Of course, many academics had the privilege of being in relatively stable jobs that could be done remotely, and many had at least some of the needed digital devices and Internet connections at home to make the transition manageable. Consequently, ours was not nearly as dire a situation as numerous people with considerably fewer resources and less stability. Still, when lockdown measures took effect, we scrambled, like so many others, to figure out our home office situation while keeping tabs on faraway family members. I personally had not had a home office setup since I had been in graduate school; my dining room table became my work headquarters.

It is hard to think back to our mindset in March and April of 2020 given everything that has happened since, but it is instructive to remember that back then no one had any idea just how long these circumstances would last. Indeed, as of this writing in 2021, we still do not. It is almost laughable in retrospect, and certainly hard to appreciate, but it is worth explicitly remembering that most places locked down with the intention of staying closed for a mere number of weeks. Writing this over a year into working from home, our naivety in those early days is cringeworthy. It is worth noting, however, that thinking the lockdowns may only last a few weeks added a sense of urgency to research looking to capture people's immediate experiences of this new situation. Put another way, we did not know how generous a window we had to construct a survey and gather data from people to capture their experiences under lockdowns.

Granted, even with hindsight knowing that lockdowns would continue on, I argue that capturing the reality of those initial weeks and months has its unique merits. Everybody was in a great unknown and seeing how people were able to use digital technologies to cope with those unparalleled circumstances sheds light on who got to benefit from better access and skills. Just as recalling our emotional state of mind from that time is nearly impossible without daily diary entries (which in retrospect I so wish I had kept), recapturing people's experiences of initial pandemic responses would be impossible in retrospect. Accordingly, a major contribution of this project is to have captured those initial impressions and experiences in real time. Beyond the historical relevance, this is also significant because people were dying of this horrible disease and, as the book shows, what information people had about it was very much linked to whether they took precautions to stay safe with implications for their health.

At first, it seemed like my research team would opt for focusing all of our efforts on figuring out how to function under lockdown. I put out some feelers to see if team members were interested in collaborating on a study, but recognizing everyone's difficult situation, I made it clear that this was 100% optional. I knew it was not something I wanted to jump into on my own, though; it would simply be too much to take on single-handed. Reactions from my team were lukewarm at first and it looked like we would sit this one out. Regardless, we were frequently in touch as I felt it important to maintain regular communication so that no one (including me) would feel too isolated.

Continued exchanges a few days later, however, suggested interest from some team members in making the pandemic a focus of scientific inquiry. After all, as communication scholars, much of society's reaction to the pandemic was directly related to our research expertise. Suddenly everyone in the wider world was interested in how people were using digital media to connect with family, friends, and colleagues. These were topics we had been studying for years. Now that they were taking center stage for everybody, how could we sit this out? It felt like we needed to be part of the conversation. We could not lend a hand in the medical realm, but we could contribute as social scientists to understanding how people were coping. It is worth noting here that those of us involved in these initial stages of the study had no family caretaking responsibilities at the time. The fact that we managed to pull off this work reflects those particular circumstances.

On March 24, we decided to launch into the project, which would result in a fifteen-minute draft survey a week later and data collected on 1,374 American adults by April 8. Ten days later, we had also launched the survey in Italy and Switzerland, taking extra time as I secured additional funding and we coordinated the translation of the instrument to three further languages (German, French, Italian) as well as making questions locally specific when needed, such as those concerning local media sources. Of note is that such speed is extremely rare; surveys usually take months of careful preparation. We had to move unusually quickly if we were to capture people's initial experiences with the pandemic, as was our goal. Fortunately, the Institutional Review Board (IRB) equivalent at the university—the Canton of Zurich ethical guidelines—is not as time-consuming to follow as most IRBs in the United States.

To coordinate our efforts, we relied on some of the same digital tools that we were studying. Fortunately, we had started using the communication platform Slack years earlier and were thus already accustomed to sharing documents remotely and communicating effectively online. We held lots of video meetings, making sure that some of them focused on social catchups, rather than work, to keep our sanity intact.

I was reminded of doing my dissertation data collection at Princeton University in New Jersey when the Twin Towers were attacked on September 11, 2001. This was significant not only because of the general chaos and malaise that followed, but also because my project was directly affected by those events. At the time, I was interviewing and observing people, some of whom it turned out had been personally affected by the tragedy. Additionally, due to anthrax, post offices across the county where I was recruiting respondents through the mail had been shut down, thwarting my recruitment efforts. This is all to say that I was not new to doing research under special circumstances and drew on those dissertation days to recognize that, while difficult, efforts under unusual conditions can yield important insights.

In mid-April, it became increasingly clear that the pandemic was not about to subside. We decided to field another survey in the United States in May, a month after the first one. By then we had had the opportunity to look at results from the first survey and knew we were onto something when it came to uncovering discrepancies in people's digital media experiences. We were also constantly coming up with new research questions that would require additional data. At this point, I gave everyone on the team the opportunity to have a section of the survey address their particular research interest so that even those who had not been actively involved in the first stage because of competing obligations could take part. In our weekly research team meetings, those on the project had been giving regular updates and so those not yet actively engaged kept abreast of what we were doing. They became inspired and were excited to join more actively once they had figured out their "new normal."

In parallel, I was also working with some colleagues from elsewhere—inconveniently in some cases with a nine-hour time difference—to craft questions. In the end, the survey got to be so long that the best course of action was to split the instrument in two. While more expensive, this was in the interest of avoiding respondent fatigue, which is an important consideration for two reasons. First, from an ethical perspective, it keeps the comfort level of the participants in mind and ensures that they are not being asked to spend too much time on the study. Second, from a data quality perspective, it avoids burdening respondents to a point that the quality of their responses ends up being questionable (see the appendix for other steps we undertook to ensure data quality). The May survey thus became two with some overlapping, but many different questions that we fielded at the same time in the United States to two different groups of people.

I share these details partly for methodological reasons, but mostly to convey the unusual circumstances under which our study materialized. As I describe in the introduction and the appendix, we put considerable effort into making sure that ours was a quality study. Nonetheless, administering several surveys cross-nationally at the height of extraordinary global mobility restrictions imposed limits on what we could do partly because of what we ourselves could accomplish under such circumstances as we pivoted to life under lockdown, but also as a result of what services were available for administering studies.

Documenting people's digital experiences during the early days of the pandemic was important for reasons I elaborate on in the introduction. Hindsight always suggests additional questions one could have asked, and longer-term projects can certainly probe those. We will never again have the opportunity to capture in real time the initial days of COVID-19, however. I am proud of what we accomplished, given the circumstances, and believe it constitutes a unique contribution to understanding how digital media played a role in what people were experiencing in these exceptional times.

The goals of this book are threefold. The first is to document people's digital media experiences during lockdown, especially as these pertain to communicating and learning about the pandemic. Second is to show how these experiences varied by societal positions. And the third is to link digital contexts and online behavior to life outcomes—namely, being informed about the pandemic, avoiding misconceptions about it, and staying safe amid a quickly spreading deadly virus. The word *unprecedented* may have gotten overused in 2020, but the reality is that switching life to lockdown was indeed an unprecedented experience for most people across the globe. Given this unusual situation, there is value in documenting how people approached these unsettled times in the realm of the digital in the immediate aftermath of having to shelter in place and what this says about variations in how people from different backgrounds incorporate digital media into their lives.

Those used to having twenty-four-hour high-quality Internet access may think that needed technologies were at the ready for everyone, thereby allowing a quick pivot of their day-to-day activities to a mediated environment. However, the book shows that this was by no means a universal experience, whether considering within- or cross-country variations. By highlighting these differences in digital circumstances, the discussion cautions against assumptions about widespread ubiquitous connectivity and offers guidance for how to avoid the inequalities that the pandemic perpetuated. While anchored in digital inequality scholarship, the findings have important implications for several areas of research, including computer-mediated communication, information seeking, news consumption, and investigations into the spread of both knowledge and misinformation beliefs. The latter two, although certainly related, turn out not simply to be two sides of the same coin.

Digital inequality refers to the differences among people who have crossed the digital divide from nonusers to users.[1] The oft-used phrase *digital divide* refers to the haves and have-nots of the Internet age. While some people remain entirely off-line, in advanced economies the vast majority of the population is now online.[2] But this does not mean that everybody is equally connected. Digital inequality captures both the variations in people's online experiences and how these replicate traditional markers of social inequality such as disparities by education and income. How sociodemographic differences translate into diverse uses and how people experienced these during the disastrous spring months of 2020 are the focus of this book.

The COVID-19 pandemic upended everyday life all over the world. Stay-at-home orders and business closings imposed a new reality on many. With mobility and public activities greatly curtailed, populations in many countries experienced a restriction of activities and social contacts unseen in their lifetimes. In order to deal with this disruption of everyday life, people turned to online resources, expanding their digital activities and—in many cases out of necessity—learning new ones. While having to forgo meetups with friends, in-person schooling, haircuts, and so many other taken-for-granted daily activities, people ramped up online purchases, participated in remote work, joined video conferences, and spent considerable time on social media.[3]

For large portions of the population, the Internet became an essential lifeline. Trying to understand what in-person activities were still allowed; figuring out how to meet basic needs of accessing medications, groceries, and toiletries; learning about the virus and preventive measures; and providing support for loved ones near and far were just some of the crucial tasks that people were now mainly doing online. Considering the extent to which people from varied backgrounds were able to take advantage of digital media in these essential efforts is the focus of this work.

This book documents how long-existing social and digital inequalities played a critical role in the extent to which people were able to pivot to much of life happening through computer-mediated communication. Regarding traditional markers of inequality, it shows that those more highly educated and with more resources did more online than their less privileged counterparts. In the realm of the digital, it demonstrates that people's abilities to use the Internet as well as their access quality was a significant factor in the extent to which the Internet truly was a lifeline. In particular, as the ultimate real-life outcome, it explores how being informed about the pandemic while steering clear of misinformation related to avoiding risky behaviors by following stay-at-home guidelines.

Depending on the Internet for a myriad of tasks was certainly nothing new in 2020. Many people were already relying on and benefiting from digital media for numerous everyday tasks. What made the pandemic events unprecedented in the realm of the digital is that it was the first time in the Internet's history that it became completely front-and-center for the digitally connected parts of the world that were suddenly relying on it for the most essential of daily needs. Even for the few allowable activities such as going to the pharmacy and procuring groceries, those with the necessary resources—both in financial terms and know-how—would have been able to opt out of such trips more easily by pivoting to ordering everything online to avoid contact with others. What did it mean to rely so heavily on digital resources that had never been equally distributed across the population? How might people be advantaged or, instead, hampered by their prior online experiences and digital skills?

For two decades, literature on digital inequality has shown that even once people get connected, moving from the have-nots to the haves side of the digital divide, differences remain in how they incorporate digital media into their lives, resulting in a spectrum of opportunities as well as disadvantages.[4] From variations in device ownership and data access quality and quantity, from disparities in skills and available social support for tackling questions about digital media uses to what people do once they are connected, how people incorporate the Internet into their lives differs considerably. How these differences played out during COVID-19 lockdowns is the contribution this book makes to our understanding of spring 2020 events.

The sudden extreme reliance across society on digital media could hinder people who did not have the necessary infrastructure in place (such as reliable quality connectivity and device access) and who lacked the needed skills to shift the majority of their daily activities to digital tools and mediated interactions. It could also put in harm's way those who did not know how to find, where to find, and how to evaluate crucial information about the virus, which was essential for avoiding exposure to it.

Accessing and understanding information about the novel coronavirus could mean the difference between health and illness, or worse, death. Rarely has fast and widespread understanding of information been so vital, yet people vary in both their abilities to locate content online and to evaluate its credibility.[5] Given the spread of considerable misinformation related to COVID-19 already in the early days of the pandemic, widespread knowledge about precautions was never a given.[6] Investigating who was most likely to believe misinformation and how this linked to Internet uses and skills—as well as the use of more traditional media—is crucial for preventing such problematic outcomes in the future.

FROM DIGITAL DIVIDE TO DIGITAL INEQUALITY

In the mid-1990s, as the Internet started diffusing to the general population in certain parts of the globe, soon concerns arose about its unequal spread. The phrase *digital divide* came to characterize these inequalities first made evident through a report by the US National Telecommunications and Information Administration titled "Falling Through the Net,"[7] which highlighted differences in connectivity by age, gender, race/ethnicity, education, income, and urban-rural residence.

The approach in those early days of investigating the digital divide focused very much on whether a household had access to the technology rather than measuring use at the individual level.[8] The report noted:

> At the core of U.S. telecommunications policy is the goal of "universal service"—the idea that all Americans should have access to affordable telephone service. The most commonly used measure of the nation's success in achieving universal service is "telephone penetration"—the percentage of all U.S. households that have a telephone on-premises.

This focus on the telephone was relevant because most households at the time accessed the Internet through a dial-up modem. In the following years, many academic studies and other reports followed, continuing to document basic access differences.[9] Notably, however, scholars also started to call attention to the fact that even after people gained access to the Internet, numerous differences remained among them.[10]

A focus on *whether households* had access to the infrastructure would ignore important variations in *how individuals* would be using the technology. Moreover, examining the technical side of access only, while ignoring the social side, would miss a major piece of the inequality puzzle. In addition to needing to recognize more nuanced aspects of the technology, such as variations in the quality of hardware, software, and connectivity speeds, other aspects also merited attention, such as ease of access to devices (not restricted to home access only), differences in available support for when people ran into problems, differences in skills to use the technology effectively and efficiently, and differences in what people did once they were connected.[11]

To emphasize the nonbinary nature of inequality related to Internet access and usage, my graduate advisor Paul DiMaggio and I proposed the term *digital inequality* in a report published as a working paper of Princeton University's Center for Arts and Cultural Policy Studies in 2001. It is worth noting that the report was rejected for publication in a journal special issue focusing on the digital divide suggesting that the arguments it was making were not widely accepted at the time. It is also worth noting, however, that the report has been cited more than 1,500 times in the past twenty years, including over one hundred times in 2021, indicating that although it may have been a bit ahead of its time, it captured an important aspect of the unequal ways in which digital media were diffusing across the population.

Over the past two decades, thousands of papers have been written about digital inequality spanning all corners of the globe, although the majority has come out of North America and Europe. Such work consistently finds variations in how people adopt and use digital technologies. That is, even once people become users, they vary significantly in what they end up doing online from the services they adopt (e.g., which social media platforms they join [see chapter 3]) to how actively they engage, if at all, in different online communities. Importantly, research has consistently documented that these variations are often related to people's socioeconomic status with those from more privileged backgrounds engaging in more diverse types of online activities than their less privileged counterparts. One significant factor in usage differences concerns people's ability to use digital media effectively and efficiently, what the literature interchangeably refers to as "Internet skills," "digital skills," and "digital literacy."

DIGITAL SKILLS AND WHY THEY MATTER

Digital skills encompass knowing what is possible through digital technologies and the ability to engage with those possibilities effectively and efficiently. Such skills have multiple dimensions, from awareness of what can be done online to how to engage on various platforms in ways that benefit the user and that help avoid negative outcomes such as scams.[12] They often have both social[13] and technical[14] dimensions because they not only concern where to click on the screen but also social considerations of what type of communication is most appropriate depending on the circumstances of who is communicating what information to whom in what particular context.

Awareness of what is possible is a prerequisite for many online actions such as setting up a social media account in a way that respects one's preferred privacy specifications. People who do not realize that a service allows users to make choices about what data it collects will not know to set up alternatives to the default settings. While most people tend to stick to default settings, this is not necessarily a reflection of their preferences, rather, it also reflects their awareness—or lack thereof—of the option to change said settings.[15]

Similarly, people who do not know that apps on phones offer varying notification options will not know to change the types of notifications they receive. Such seemingly minute details matter, because they concern people's everyday experiences with technologies that they use constantly. Research on disconnection from various technologies and services shows that one reason people abandon apps is that they feel overwhelmed by the amount of information coming at them.[16] Notifications can be a significant culprit in this, whether through apps on one's phone or through endless email messages. Were people to know that it is possible to turn off notifications, their impressions of and feelings about certain services may change. Of course, knowing that something is possible is not the same as knowing how to accomplish it. Even if someone realizes that they can turn off notifications, it is not a given that they know how to do it. This is where the many other dimensions of skills come into play.

To communicate effectively in a mediated environment, it is important for users to choose the communication functions and capabilities most appropriate for their goals. Just like in face-to-face situations, the context of one's interaction gives cues to what types of communication styles are most appropriate. For example, there are different expectations for how two coworkers communicate with each other compared to how two best friends share information.[17] In the online environment, such societal roles continue to influence what are appropriate forms for interaction, but rather than the one in-person context for communication, there are numerous channels through which the interaction may proceed. In addition to the style of communication (e.g., the use of formal versus informal language), what platform or medium one uses to reach another is also a relevant consideration.[18]

Although it may be appropriate to send a text message to a friend, in many contexts it is more appropriate to send an email to professional contacts.

An especially relevant domain for varying levels of digital skills is information seeking and credibility assessment.[19] Decades of scholarship has investigated how people find content online and decide whether it is something they should believe and trust. Indeed, some of the earliest work on Internet skills was specifically preoccupied with how effectively and efficiently people could find certain types of content and evaluate it.[20] This research established that those from more privileged backgrounds did better at locating relevant materials proficiently than others.

Other domains of digital skills include the ability to use specific platforms like social network sites (e.g., Facebook, Instagram, Twitter), share one's own content in ways that it reaches desired audiences, know how to protect one's privacy and security, deal with the onslaught of online information, and manage one's online reputation in optimal ways.[21] The mix of these skills allows users to make the most of their time online from setting up services in ways that meet their needs to finding information they seek, evaluating it for what it is, sharing it if they so wish, and interacting about it when they are so inclined.

A notable finding of research on Internet skills is that those who have more of it tend to engage in capital-enhancing activities more than those who are less skilled. Capital-enhancing activities refer to types of Internet uses from which one may benefit. This can include following the news to become a more informed citizen, taking advantage of government services to receive needed assistance, finding commercial deals to accrue savings, network for improved job opportunities, learn about preventive medicine for better health, and the list goes on.[22]

What constitutes capital-enhancing uses may depend on individual circumstances, however. Closely following the weather may just be a hobby to some, but for farmers it has important implications. Browsing do-it-yourself videos may be a recreational activity for one person while it provides job-related skills training for another. Existing digital inequality research often groups online activities into categories like "capital-enhancing" versus "recreational" or "entertainment-based," but some have argued that engaging in leisure online can be significant in its own right.[23]

One particular type of online activity that has gotten less attention from digital inequality scholars concerns actions with potentially harmful outcomes such as falling for online scams, believing in misinformation, or being the victim of a phishing attack (the process where others try to gain unauthorized access to someone's account by tricking them into sharing their password).[24] Although there is considerable research on such experiences more generally (e.g., falling for scams, believing misinformation), such work rarely looks at how actions that result in problematic outcomes vary by socioeconomic status and other markers of inequality.[25] One such problematic outcome concerns misinformation beliefs about a deadly virus. This book explores who is most likely to believe such misconceptions and what activities result from holding them.

WHY DOES INTERNET USE MATTER WHEN PEOPLE'S
LIVES ARE ON THE LINE?

With more than 6 million people having already died by spring 2022 across the globe (over one million in the three countries that are the focus of this book) and countless others surviving only on ventilators or in a coma, bedridden with serious long-haul conditions, is it warranted to focus an entire volume on people's digital media uses at the time of initial lockdowns? It is, for multiple reasons. When it was no longer possible to go to places physically, most interactions, whether professional or personal, moved to mediated communication channels. Having the opportunity and ability to pivot to performing daily activities online became crucial in maintaining links to the outside world. Who got to make that transition smoothly has important implications for how people managed in these unsettled times.

For staying abreast of essential information about the pandemic, Internet use could mean a big difference. While traditional media like newspapers, radio, and television were certainly an option for getting information about the pandemic, having the opportunity to seek out locally relevant news at a time when place-based traditional media had been seriously curtailed could be nontrivial for many.[26] Access to such information was imperative for appreciating the state of community affairs including guidelines for what to do in one's geographical region and, importantly, what not to do. Following the spread of the virus globally could be interesting, nationally it could be helpful, but especially in one's immediate surroundings, it could be lifesaving.

With the closure of businesses, the livelihood of many people was cut off, resulting in sudden massive unemployment. For people impacted in this way, finding alternative sources of income was, again, significantly dependent on the ability to navigate online resources. Although freelance jobs and various microtasking options may well be available online and many do not depend on being in physical proximity of an employer, knowing that such opportunities exist, finding the most relevant ones, being able to register for them and meet the service's requirements, and having the necessary hardware, software, and connection quality to fulfill the services are all a precondition to being able to benefit from them. Walking into a business to inquire about job openings was an unlikely prospect; rather, pivoting to do the online equivalent could mean the difference between an empty bank account and an income.

Similarly, navigating government assistance requires being savvy with technology, both to find out about and to apply for relief. Although contacting services over the phone may sometimes be an option, confusing phone trees and indefinite wait times can make such alternatives difficult and frustrating. Again, having the know-how to find relevant forms, print out, sign, scan, and upload documents as needed all require varying types of resources that are not equally distributed across the population. With libraries also closed to patrons, those traditionally dependent on their services for help with computers and access to printing were suddenly cut off from their customary means of conducting such logistical tasks.

The rise in online shopping is certainly not a novelty of the pandemic, but few have depended on it for everything in their lives, including such basic necessities as groceries and medications. Certainly, many had yet to embrace e-commerce by 2020 for varying reasons ranging from not knowing how to navigate the process of online shopping to security concerns about credit card numbers being stolen and the threat of identity theft. But during lockdown, from one day to the next, there were suddenly few alternatives. And while grocery shopping remained an option even in the most locked-down of places, going outside posed a risk, and so when possible many would have presumably preferred to avoid it.

Knowing how to access online shopping options may seem trivial to those who had done it for years, but it is not necessarily an obvious undertaking for those new to the experience. Such financial transactions also tend to require a credit card or other online arrangements, access to which not everybody can take for granted. Accordingly, online shopping as an alternative form of procuring basic necessities was more within the reach of some than others.

Digital access to numerous other services also cannot be taken for granted, even among those who were already regular users of certain technologies. Getting medical advice, receiving therapeutic services, maintaining social connections—all are possible online, but require know-how. Last, but certainly not least, digital media offered important outlets and inspiration for unwinding. Whether access to movies at the click of a button or tips for the perfect sourdough starter, having the means and ability to connect around recreational activities could also be important for mental health, but itself required familiarity with and access to available services and communities.

Given these many ways in which digital media can improve people's situations in difficult circumstances, there is much value in seeing whether it met this need, and if yes, whether and to what extent these benefits accrued to people across society or more so to the relatively privileged. Put another way, do digital media improve people's circumstances, and if yes, does this happen at similar rates across diverse population segments or do some gain value more than others? Figure I.1 shows a graphical depiction of the process proposed by the theory of digital inequality about how digital context and use is influenced by user background and then may in turn influence life outcomes.

Sociodemographic background such as age, gender, education, and income have the potential to limit or boost what types of hardware, software, connectivity speeds, and data limits people have at their disposal—especially at home—what types of support are available to them, and their level of digital skills. This digital context then is likely to influence what people do online and how much time they spend using digital media, behaviors also

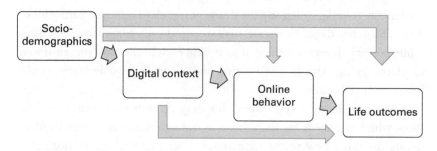

Figure I.1
The theory of digital inequality: the process by which differences in social position translate into varied digital contexts, which influence online behavior and may result in divergent life outcomes.

related to their background. All of these factors then have the potential to influence life outcomes, which in the case of a global health crisis includes understanding how the virus is transmitted, how to stay safe from it, what to do when infected, and how to avoid misinformation about it. Thus, well beyond whether someone is an Internet user or not (the traditional understanding of the so-called digital divide), how people incorporate digital media into their lives once already connected is both a result of and may result in their particular societal position.

A COMPARATIVE PERSPECTIVE: THE UNITED STATES, ITALY, AND SWITZERLAND

While investigating the digital aspects of how people coped with the pandemic in just one country would already provide plenty of material for analysis, I opted to make this a multicountry study. Why complicate things by comparing countries when just one offers plenty of material to explore how people were dealing with these unsettled times? The value of comparison cases is that they can help put findings into perspective. When results generalize across cases, it gives more confidence in those findings. When they do not, it can raise interesting questions about what local factors may explain the variations and thereby shed light on what is ultimately driving particular findings.

Why the cases of the United States, Italy, and Switzerland? They are all Western democratic countries that are relatively well off, so their comparison controls for certain structural advantages in the global context. Especially relevant to a study of digital inequality, they are all relatively well wired with the majority of their populations already Internet users in prepandemic times. They differ in important respects, however. Their health care systems vary, which may be relevant in a pandemic. In the United States, health insurance continues to be tied to employment in significant ways and otherwise is left up to the individual. During a health crisis when many people's jobs are at stake, health insurance being contingent on one's employment can cause additional strain on those who are already in precarious positions. Italy is part of the European Union and has a national health service that provides universal coverage with public health care largely free of charge. Switzerland has a privatized health insurance system; there is no free state-provided health care, but health insurance is mandatory for all citizens and residents and is not tied to employment.

The three countries also have very different social safety nets in place, which is relevant in a situation where many people's employment was at risk. The media systems also differ given that Italy and Switzerland have strong public broadcasting systems, whereas such institutions are much weaker in the United States. It is not the goal of this book to analyze the structural conditions of the pandemic, but having state-level differences can make for interesting comparisons. Hopefully other scholars who focus on that level of analysis will offer their own lessons learned.

Italy in particular was a relevant case to study in April 2020 because it was the first country outside of Asia to be affected by COVID-19 in a major way and thus had more experiences with it than did others at the time. This may seem naive in retrospect, but it was not clear in March 2020 that so many parts of the globe would be significantly affected by the virus and for how long. After all, other viruses of recent times such as H1N1 and Zika had been much more geographically contained. Accordingly, our research team's general sentiment was that if we wanted to capture people's experiences of lockdowns, we had to collect data quickly and doing so in an area that had been especially hard hit seemed strategically advisable to capture people's lived experiences with lockdowns.

While the urgency of data collection may not seem as obvious in retrospect knowing just how long the pandemic lasted, there is nonetheless great value in having captured the initial experiences in some of the places that were hardest hit early on. Italy had a hard lockdown unlike Switzerland's soft lockdown. It also borders Switzerland with considerable traffic between the two on a daily basis and indeed, the first Swiss case was reported in the Italian-language part of the country. Chapter 1 provides more details on these local pandemic conditions.

Being based in Switzerland, studying the situation from there was an obvious choice for my team, with the hopes of understanding the local situation better. Including the United States was meaningful because many studies of digital media usage—including many of my own—have been administered in that country, so there would be helpful baseline comparisons in the literature to assess which findings reflected the norm and which were unique to the present situation. While translating surveys to different languages and adapting questions to different national contexts takes considerable work, the ability to make cross-national comparisons is well worth it.

Given the large geographical expanse of the United States and that it has one of the largest income inequalities among members of the Organization for Economic Co-operation and Development, it is perhaps not surprising that variations continue to exist in who is online.[27] According to the Pew Research Center, people in rural areas are less likely to have broadband (i.e., high-speed) access than others: 72% compared to 77% of urban and 79% of suburban Americans.[28] While 90% of Americans were online in 2019 (with 93% in 2021), this varied by user background such that 84% of people with no more than a high school degree were Internet users compared to 98% of those with a college education. Age continues to be a strong correlate of Internet adoption with about three-quarters of those sixty-five or older online compared to almost everybody in other age groups.

Italy is the least wired of the three countries. The World Bank's latest figures are for 2018, when 74% of the country was using the Internet.[29] Age is the strongest correlate of use with only 35% of those fifty or older online in 2015, a figure that is well below the European average of 49% at that time.[30] There are large variations across the country with people in the south much less likely to be connected.

Switzerland is the most wired of the three countries with 93% of the population online in 2019.[31] People ages fifty to sixty-nine at 93% connectivity are not far behind those younger than fifty, who are close to 100% online. However, among adults age seventy and older this drops to 60%, and given the large portion of Switzerland's population that is over seventy, this does make a dent in the country's connectivity statistics.[32] Chapter 2 discusses the digital context of the three countries in more detail.

METHODOLOGICAL APPROACH

Knowing how particular online activities relate to people's life situations can be difficult without asking them about it. While there have been exciting innovations in the social sciences in the domain of analyzing logs (automatically recorded details) of people's behavior, survey methods continue to offer unique insights difficult to obtain through other approaches.[33] Given the questions of interest in the project, asking people directly about their experiences of the pandemic coupled with questions about their demographic and socioeconomic background was the best available methodological option.

My team collected data from adults in the United States, Italy, and Switzerland in April 2020 and then some additional data in the United States a month later. We contracted with Cint, an online survey company. These survey responses form the basis of the empirical results discussed in this book. We surveyed 1,374 American adults age eighteen and older from all fifty states and Washington, DC, on April 4–8, 2020. We surveyed 983 Italian adults age eighteen and older across the country on April 16–17, and 1,350 Swiss adults age eighteen and older across all twenty-six cantons on April 17–24. In May, we surveyed an additional 1,551 Americans. Analyses of the latter data set make up a small part of the book, but are helpful for addressing a few questions that were not available in the earlier survey. The appendix offers more details about the survey methodology including what safeguards we implemented to ensure data quality. It also explains how I applied weights to the data sets to adjust for discrepancies compared to the general population. The appendix also gives an overview of the basic sociodemographic characteristics of the different samples.

Given that we used online survey methods to gather information about people's experiences of the pandemic in near real time, an important caveat is in order specific to the focus of this book. By definition, those who are at the lowest end of the connectivity spectrum either because they are not at all online or only have spotty access and low user skills are unlikely to be in the population that participates in online surveys. Accordingly, such people will be largely absent from this book. Put another way, there is some sampling on the dependent variable in that participation in the study required some level of connectivity and digital savvy in the first place. This is unfortunate, but a necessary limitation in a time when little could happen through other means, especially at the speed at which it was important to gather data in order to capture the in-the-moment experiences of the pandemic's early stages. It is notable, nonetheless, that there are variations in access and usage—some mobile-only users, some infrequent users—even in this group, making it possible to identify important differences and showing that the sample is diverse not just in sociodemographics but in online experiences as well.

The implication of the online-only survey mode is that the findings in this book are likely conservative when it comes to identifying the contours of digital inequality that both affected and were brought to the fore through the pandemic. That is, when examining how many people had only one device with which to access the Internet at home, the results here will not generalize to the population because those with limited access are less likely to participate in an online survey and thus their proportion here will be lower than in the population. Because the book finds both device access and skills to make a difference in how people benefited from digital media during the pandemic, including even more people who fare poorly on those accounts would likely make the divisions even more pronounced. To that end, differences identified here likely underestimate the true extent of digital inequalities in each country studied here.

Relying on respondents who have the time and inclination to take a survey poses other notable limitations. Those who were most overwhelmed by caregiving responsibilities, by the need to find new employment, or who were themselves infected by the virus and suffering in serious ways are also less likely to show up in such a study. This is less related to the online mode of data collection and more to the fact that it took time to answer questions. Much more unobtrusive methods would be needed to capture data about such people, but alternative methods often make it impossible to get at some of the questions of interest here and pose their own set of limitations.[34]

An important question across the chapters of this book is whether sociodemographic characteristics mattered for how people experienced the pandemic in the realm of digital technologies. That is, did people of different education levels, for example, experience the digital side of the pandemic differently? Of most interest are sociodemographic characteristics that have been known to matter for questions of digital inequality such as age, education, income, and disability status, although others such as gender and geographical residence will also feature in some of the discussions.[35] In the US context, race and ethnicity are also important to examine and are included in the analyses. While 2020 saw significant mobilization around the Black Lives Matter movement, this topic is not a central focus of this book as the data collection was already over before George Floyd's tragic death on May 25 and the events it precipitated. (I will note here that I capitalize all racial identifiers following the lead of the National Association of Black Journalists and others.[36])

To examine whether experiences vary by user background, I rely on different statistical analyses. In some situations, I report on simple bivariate analyses in which I only look at the relationship of two variables. An example of this is to compare the digital skills of people of different ages (figures 2.3 and 2.4). In most cases, however, I will go a step further to report on the results of more advanced statistical analyses that take into account other factors. Following the same example, it may be that younger people and older people differ in their skills, because younger people spend more time online and thus differences in skills reflect varied online experiences rather than varied ages. I will discuss such findings of analyses that take other factors into account by using phrases such as "controlling for other factors" and "holding other factors constant."

Also worth highlighting is that I will only claim that differences exist when these are statistically significant at levels generally agreed on to be notable: at the 95% confidence interval. This means that there is only a 5% chance that the differences observed in the data set are due to random chance.

Although there were numerous data collection initiatives in the first weeks of the pandemic, not all of them relied on national samples. Those that did concentrated on a myriad of topics from health issues to job loss. This project is unique in its focus on the digital aspects of the pandemic, arguably a crucial aspect of how people were experiencing the changes brought about by lockdowns and the need to keep safe amid a quickly spreading deadly virus. It is this feature of the data that makes this book possible, and it is because of this particular focus that this project is distinctive in the large landscape of studies about COVID-19. Accordingly, the surveys we conducted are the main source of data for this book. While we put great effort into ensuring quality, the topical coverage of our questions was restricted not only by resource limitations (e.g., number of items we could ask), but also by how far our own knowledge and imaginations could go to come up with relevant questions to ask within the short window of time from the start of lockdowns to data collection. Fortunately, other scholarly initiatives followed. For a sense of people's digital experiences during initial lockdowns, this text is unique.

CHAPTER STRUCTURE

The preface described the context of such an in-the-moment study and this introduction gives background for why examining lockdowns through a digital inequality lens is important. Chapter 1 will give social context about the pandemic in the three countries examined here. In addition to some descriptions of how people were coping in the initial days of lockdowns, importantly, it describes people's knowledge and misconceptions about COVID-19. As later analyses will show, understanding versus holding misbeliefs about the virus was directly linked with whether people abided by stay-at-home guidelines. Because not taking those seriously could have dire health implications, it is a crucial outcome to examine. Chapter 1 sets the stage for what people did and did not understand about the virus and how sociodemographic background related to their understanding.

Chapter 2 drills down on the varied digital contexts through which people experienced the early weeks of the pandemic. While some had multiple points of access at home and could switch to remote work with relative ease, others scrambled to find a footing, resulting in worries about having Internet access. People also varied in the availability of social support they could tap to help figure out technology-related questions. On the whole, digital circumstances mirrored social advantages with those from more privileged positions enjoying more types of access. More robust device ownership also translated to higher Internet skills. Importantly, both better access and higher digital skills are positively linked with more knowledge about virus risks and prevention efforts even after taking sociodemographic differences into account, showing why digital inequality matters.

Social media are some of the most popular online services and chapter 3 focuses on how people were using such apps and sites to connect about COVID-19. The chapter first establishes that the different platforms attract people of different backgrounds, a point worth recognizing because it has implications for whose content circulates on them and who sees what materials. Next follows a discussion of the types of pandemic-related content people saw, shared, and discussed on social media, showing some variations across the three countries and by user background. The chapter ends by illustrating how such activities related to feelings of social connectedness as well as knowledge and misconceptions about the virus.

A key aspect of spring 2020 was widespread confusion about the novel coronavirus, including such essential information as how it spread and how this could be prevented. Chapter 4 looks at the larger landscape of information sources people used to keep abreast of the pandemic from traditional media like television and newspapers to online sources such as news, health, and government websites. Television, in particular, proved to be a major source of content for people across the three countries, and their national broadcasters (the major private networks in the United States, and the public broadcasters in Italy and Switzerland) played the most important role in disseminating information that helped people's knowledge, and in the case of Switzerland, avoid misinformation.

Finally, the conclusion returns to the model of digital inequality described earlier in this introduction to show how people's varied social and digital circumstances translated into differentiated online behaviors concerning the pandemic, which then ultimately linked to how much they abided by stay-at-home orders.

BEYOND THE CASE OF A GLOBAL PANDEMIC

The story that will unfold in the following pages is one of differentiated digital contexts linked to varied online experiences with real-world consequences concerning what knowledge and misconceptions people held about the novel coronavirus. Vast bodies of scholarship exist on both how people gain knowledge through information sources and the disconcerting spread of misinformation. While the former has been linked to questions of inequality for decades through substantial work on the knowledge gap hypothesis, which I discuss in chapter 4, the latter does not have a long history of being considered by markers of social stratification. This book connects digital inequality scholarship with misinformation research to suggest that there are fruitful ways in which both can benefit from the other.

Another contribution of the book that goes beyond COVID-19 is the concurrent examination of both knowledge acquisition and misinformation beliefs. Most investigations focus on either one or the other; the scholarly literature on knowledge gaps is rarely in direct conversation with work on misinformation and disinformation. One possible reason for this lack of concurrence could be an implicit assumption that more of one automatically means less of the other. While this may hold for certain types of information, in the context of a sudden health crisis, the two need not be entirely linked. It is possible to know certain facts about a virus while also holding misconceptions about it. Indeed, chapter 1 shows that while COVID-19 knowledge and misconceptions are somewhat related (negatively, as one would expect), they are by no means simply two sides of the same coin. To this end, the book offers motivation for those studying either knowledge gaps or misinformation to incorporate the other in their work as well.

While we can only hope that a global pandemic the scale of the novel coronavirus in 2020 will not be a reoccurring event in our lifetime, smaller epidemics, natural disasters, and political upheavals all have the potential to cause similar disruptions even if at a more local scale. The H1N1 virus in 2009 in Asia, the Ebola outbreak in the mid-2010s in Africa, the Zika virus in 2015 in South America were much smaller than the COVID-19 pandemic, but certainly affected large regions with millions of people in their own right. From earthquakes to floods, from tornadoes to tsunamis, from hurricanes to heat waves and wildfires, people across all continents have experienced major natural disasters that have severely limited mobility and disrupted everyday life. Likewise, violent political events continue to sweep across countries in every corner of the world. In early 2022, Russia's war on Ukraine again showed the important role digital media can play, from connecting refugees to vital resources to disseminating both informative and misleading content. Such events all come with challenges for addressing everyday needs, including one of the most basic ones—communication.

Although the primary focus of this book is the early months of the 2020 global pandemic, the lessons learned about how people from different backgrounds experience a major disruption in day-to-day activities have implications well beyond its specifics. While COVID-19 will continue to pose ongoing challenges and long-term consequences, understanding the interplay of digital inequality and people's experiences in the early days of major life disruptions is significant both for documenting such an acute time in history and for offering lessons to be more prepared in case of future such interruptions to life as we know it. Understanding how people's communication about and knowledge of the pandemic played out in these unsettled times can offer guidance for government agencies, news organizations, and others needing to reach the population with essential, rapidly changing information in a timely and effective manner.

THE SOCIAL CONTEXT OF LIFE DURING LOCKDOWN

Before diving deep into how people used digital media in 2020 to manage life under lockdown, this chapter sets the stage for how people were coping more broadly during the pandemic's early weeks. After giving a quick time line of events in the three countries explored in this book, the discussion moves on to examining people's main worries, changes in their home circumstances including childcare responsibilities when applicable, and changes in employment. Then it considers people's confidence in institutions such as the medical system and the federal government to handle the pandemic. The final section looks at people's knowledge of COVID-19 and what misconceptions they had about it, which will feature prominently throughout the book in relation to people's digital media uses.

Although implemented to varying degrees, most governments and public health institutions worldwide set physical distancing and stay-at-home guidelines to prevent the spread of the novel coronavirus in spring 2020.[1] Places where individuals typically gather such as the workplace, schools, shops, restaurants, bars, cafés, cultural venues, houses of worship, sports establishments, and beauty salons were often closed to minimize public health threats. The extraordinary stay-at-home circumstances made for fewer opportunities to meet in person, and thus many of the ways in which people connect socially were no longer available. Indeed, analysis of cell phone data showed substantial increases in physical distancing among Americans during the first months of the pandemic.[2]

Italy was the first country outside of Asia to go into lockdown. Although research late in 2020 suggested that the new coronavirus was already circulating in the country in fall 2019, it was not until 2020 that people were diagnosed with it.[3] The first confirmed cases in Italy occurred in January when two Chinese tourists tested positive in Rome.[4] By the end of February, several small northern Italian towns saw outbreaks with the first death occurring on February 22. Increasing restrictions were placed on the region followed by nationwide lockdowns on March 9.

Lockdowns in Italy were more severe than those implemented in many other countries in the weeks and months to follow. For example, going outside for a run could only be done near one's residence and people were not allowed to travel around the country at all. As the first hot spot of cases outside of China, Italy received much attention at the time and makes for an interesting case of lockdown circumstances both for being the first country in the West to experience it and due to the relative severity of restrictions put on people's mobility there.

The southern part of Switzerland borders northern Italy, precisely the part of that country that was hardest hit by COVID-19 from the start. Cases started rising quickly in Switzerland's Italian-speaking border regions due to many connections between those neighboring areas, such as workers crossing from one to the other as part of their daily commutes. Indeed, the first documented case in Switzerland occurred in the Italian-speaking canton of Ticino on February 24, 2020, with a seventy-year-old man who had contracted the virus while participating in a demonstration in northern Italy.[5] Several other initial cases in Switzerland were linked to movement across the Italian border.

Despite the significant rise in infections, Swiss authorities did not start limiting border crossings until March 12 and did not close borders to Italy (and other countries) entirely until March 18. As cases surpassed the 2,000 mark, accompanied by thirteen deaths, the Federal Council announced nationwide lockdowns starting March 16. While schools and many businesses closed, restrictions were never as severe as in Italy.

Infections were reported in the United States even earlier. The first confirmed case was in Washington state on January 20, 2020. A 30-year-old man had developed symptoms after returning from a trip to Wuhan, China.[6] The first death occurred in Seattle on February 29 with nineteen deaths across the United States a week later.[7] The first states to close their schools were Washington and New Mexico on March 13 when there were 2,163 cases nationwide and already forty-nine deaths had occurred. Many other states closed their schools by Monday of the following week, March 16. The first stay-at-home order came in California on March 19, with several others following suit in the coming days. New York City was another area of the country hit relatively early.

Given initial infections in cities, the more rural regions of the United States may not have been as alarmed at first. The United States is much larger in geographic distribution and population than Italy and Switzerland, so the spread of the virus was more concentrated in certain areas at first, although by the third week in March all states had several confirmed cases. There was never a nationwide stay-at-home order, such guidelines being instituted at the local or state level instead. All but nine states (eight of them with Republican governors) had such orders by April, with those nine states making up just 7.5% of the US population. A month later, just seven states (six with Republican governors) or 4.3% of the country's population had no such orders.

SHELTERING IN PLACE

At the time of this study in April 2020, most areas surveyed were under lockdown or stay-at-home guidelines. Figure 1.1 shows how seriously people abided by these measures in the three countries based on their responses to the survey. It is clear from these graphs that Italy was experiencing a

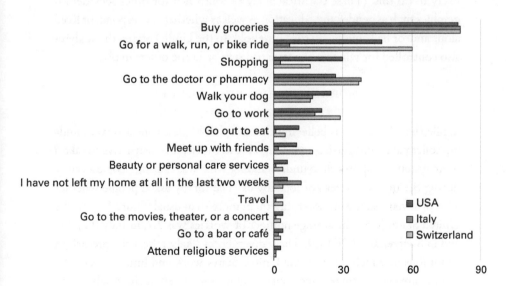

Figure 1.1
Percentage of respondents who reported having left their home for various essential and nonessential activities a few weeks into lockdown measures in the three countries.

hard lockdown compared to softer approaches in the other two nations. Although it continued to be widespread for most people to go out for buying groceries and going to the doctor or pharmacy remained relatively common in all three nations, Italians were much more limited in their mobility in most other ways. Americans were most likely to go shopping beyond groceries or eat out, while the Swiss were most likely to go to work and meet up with friends. They were also the most likely to go out for exercise. Again, the small numbers for that activity in Italy show just how serious the lockdown measures were. Indeed, 12% reported not having left their home for any activities at all, a figure that was much smaller in the United States (6%) and Switzerland (4%).

Aggregating the reasons people left home, only 4% of Italians did so for any nonessential activities (that is, meeting up with friends; going out to eat; going out for beauty or personal care services; going to a place of worship, the movies, a theater, or a concert; and going to a bar or café). Leaving the home for other-than-essential tasks was considerably more common in the other two countries: 23% in the United States and 22% in Switzerland. In all three countries, younger people and those with disabilities were more likely to do this. (These statistical analyses controlled for other sociodemographics such as gender and education as well as whether the respondent lived alone and/or had children in the household. In the United States, the analyses also controlled for whether the state had stay-at-home orders in place.)

WORRIES AND HOME EXPERIENCES

While the media had us believing that people's biggest concerns were finding toilet paper and getting sourdough bread to rise, it is instructive to take a more systematic approach to understanding what was causing people anxiety during the first few weeks of lockdowns. The survey asked people whether a list of seven issues were worrying them more than usual (figure 1.2). In the United States (US), the average number of worries was 2.7, in Italy (IT) 2.3, and in Switzerland (CH) 1.9. These divergences likely reflect the precarious conditions in which many in the United States work and limitations of the health care system there. (See "Survey Questions Used in the Analyses" in the appendix for a copy of all survey questions used in the book's analyses.)

Figure 1.2 shows what percentage of people in the three countries marked different issues as concerns. In the United States, half of respondents signaled

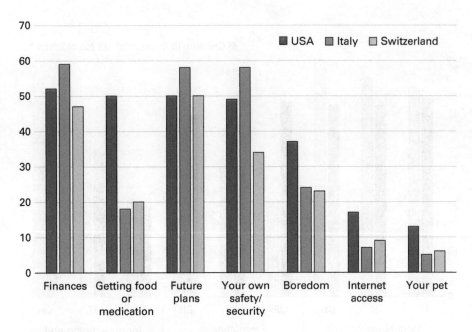

Figure 1.2
Percentage of people worrying about various issues since the novel coronavirus outbreak in the three countries.

getting food or medication as a worry, which is in striking contrast to the one-fifth of people who indicated this in Italy and Switzerland. Finances and future plans were on many people's minds in all three countries, safety and security in both the United States and Italy, less so in Switzerland. The remaining issues of boredom, Internet access, and one's pet were again more often a source of worry among Americans. About one in ten American respondents said that none of the seven issues were a source of concern; in Italy, only one in twenty marked no concerns; in Switzerland one in six (17%) were not worried about any of these issues.

When asked how the pandemic was affecting people's circumstances at home, more people listed positive ones than negative ones. Figure 1.3 shows the positive experiences in all three countries separately for people with children under 18 in the household (darker left-hand columns) and those without. A general pattern observed is that between 40% and 50% of respondents experienced more personal time, most common among Swiss adults without children. Strengthened family or partnership was more common among

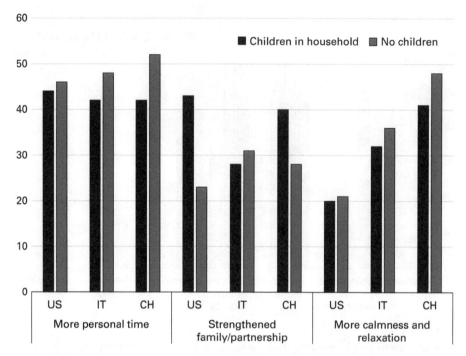

Figure 1.3
Percentage of people with positive home experiences due to the pandemic in the three
countries, separated out by whether there were children in the household.

households with children in the United States and Switzerland, but not in
Italy. While more calmness and relaxation did not seem to vary that much
by presence of children, these experiences were very different across the
three countries with those in the US reporting this the least (just one in
five respondents) compared to closer to half in Switzerland, with Italy in
between at about a third of the sample.

Negative experiences, as illustrated in figure 1.4, were generally less
common. Differences between households with and without children
under 18 are more pronounced here than with positive experiences. In all
three countries, substantially more people experienced lack of personal
space or alone time as well as tensions and conflicts in the households where
children were present. In Italy and Switzerland, there is no difference by
presence of children in whether respondents felt trapped; it was, however, a
more common sentiment among Americans with children in the household
than those without.

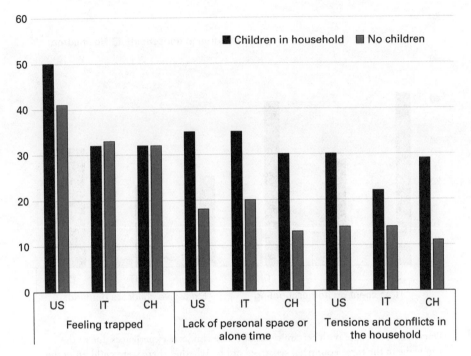

Figure 1.4
Percentage of people with negative home experiences due to the pandemic in the three countries, separated out by whether there were children in the household.

Finally, it is important to consider how the closing of schools affected those with children. Not surprisingly, few among those without minors in the household experienced repercussions. Those in that group who did may have lent a hand to friends or family members in need of childcare. Around 40% of people with children in the household reported time spent on home-schooling in the United States and Switzerland; in Italy, this was lower at 25% (see figure 1.5). There are no differences in this by gender of the respondent or other sociodemographic factors. It is likely the case that only people with children of very specific ages were particularly hit by this issue.[8] Regarding excessive childcare demands, about one-fifth of people with children under eighteen in the United States and Italy reported this and 29% in Switzerland.

The United States and Switzerland surveys also asked about childcare responsibilities among the 31% and 26% of respondents, respectively, who reported having children under age eighteen in their household. In the United States, 83% of women among these said they were at least in part

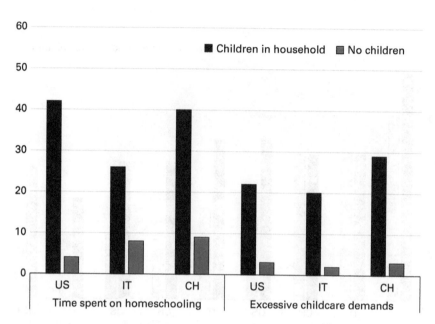

Figure 1.5
Percentage of people with certain types of home childcare experiences due to the
pandemic in the three countries, separated out by whether there were children in the
household.

providing the childcare themselves compared to 71% of men. In Switzerland, 77% of women said this compared to 64% of men. In addition to
spouses sharing in this responsibility, other family members also assisted
(16% and 12% in the two countries, respectively) while also getting some
help from friends and neighbors (4% and 6%) as well as some from childcare providers (1% and 4%). As considerable media coverage and academic
research has documented, women were often more impacted by childcare
responsibilities during lockdowns, differences that also seem mirrored here.[9]

A consistent finding across the three countries is a relationship of worries and age. In the United States, those seventy and older reported considerably fewer concerns. In Italy, those under thirty had more concerns
than adults of all other ages. In Switzerland, the difference occurred in middle
age whereby those fifty and older reported fewer worries. This is notable
because older adults were identified as an especially vulnerable group concerning COVID-19 since the early days of the pandemic. Yet this did not
seem to translate into worries per se among participants of this study.

When it comes to changes in home circumstances, there are no differences by age in positive experiences in the United States and Italy, whereas in Switzerland, those in their twenties and thirties were more likely to report such outcomes of the pandemic. As for negative home experiences, the youngest people were more likely to have them. These findings are noteworthy as some may be inclined to make assumptions about older adults having more concerns and needs while in isolation. But it is possible that older adults fared relatively well in the pandemic's early weeks thanks to more established routines around a homebound life for those already retired and not regularly leaving home for school or work.

EXPERIENCING LOCKDOWNS AS A DISABLED PERSON

Disabled people are worth distinct consideration during the pandemic given that they are a marginalized group with barriers in access to many services, including digital technologies, which the next chapter discusses. Disabled participants in this study in all three countries were about twice as likely to be in the medically high-risk category when it comes to contracting COVID-19. During lockdowns, they were sometimes targeted as less than worthy of protection, a sentiment directed at disabled people not only in pandemic times. On April 20, 2020, at a rally in Nashville, Tennessee (US), a protester was filmed holding the following sign: "Sacrifice the Weak—Reopen TN [Tennessee]." Such sentiments suggest that certain people are less worthy of life and health safety precautions than others.[10]

In Antioch, California, a city planning commissioner, since removed from his position, shared the following on Facebook: "A herd gathers it ranks, it allows the sick, the old, the injured to meet its natural course in nature."[11] He went on to expound: "then we have our other sectors such as our homeless and other people who just defile themselves by either choice or mental issues. This would run rampant through them and yes I am sorry but this would fix what is a significant burden on our Society and resources that can be used." The implication for disabled people in these statements as they pertain to their right to protections from COVID-19 could not be more clear.

It was not just random posts on social media that were a cause for concern about the well-being of disabled people, however. As hospitals were contending with scarce resources for treatment of those infected with COVID-19, disability advocates were concerned that policies favoring younger and healthier patients would leave the disabled denied the care they may need, which was indeed what happened in some cases.[12] Such challenges and threats were not restricted to disabled people in the United States. In the United Kingdom, for example, the government made it possible for local authorities to reduce the care they provided to disabled people.[13]

It is not surprising, then, that disabled people in all three countries in this study were much more likely to report worries about getting food or medication than those not disabled (61% among disabled people compared to 48% among the nondisabled in the United States). However, it is notable that this is the only domain about which disabled respondents in this study were more likely to be concerned; there is no variation in worries about finances, future plans, safety, boredom, or Internet access. They were also no less likely to have experienced the positive home circumstances included in figure 1.3 nor any more likely to have experienced negative home circumstances listed in figure 1.4 due to the pandemic compared to nondisabled people.

Disability activist Alice Wong has highlighted how efforts by the disability community to make the world a more inclusive place can benefit the population as a whole: "For many disabled, sick, and immunocompromised people like myself, we have always lived with uncertainty and are skilled in adapting to hostile circumstances in a world that was never designed for us in the first place. Want to avoid touching door handles by hitting the automatic door opener with your elbow? You can thank the Americans with Disabilities Act and the disabled people who made it happen."[14] This example shows well how accommodations developed initially with particular people in mind can ultimately serve the broader public. Similarly, ramps to assist routes with elevation changes greater than half an inch mandated by the Americans with Disabilities Act benefit adults pushing baby strollers or just about anyone pulling heavy carts or luggage.[15]

Given the particularly acute situation of disabled people in a health crisis dealing with circumstances and systems that discriminate against them, it is especially of interest to see how they used digital media during the early weeks of the pandemic. One realm in which their long-term efforts were suddenly benefiting the broader public was in the domain of remote work. Disabled people have been at the forefront of encouraging such employment options for decades, arrangements from which many well beyond that community were now benefiting as working from home became the norm in many occupations.[16] The next section briefly considers how people's job circumstances changed during the pandemic as additional context for how they were experiencing these unsettled times.

JOB LOSS AND WORKING FROM HOME

The survey asked respondents whether they had experienced any changes in their "employment status due to the Coronavirus." A notable difference across the three countries is that 7.6% in the United States responded that they had been laid off compared to 2.6% and 2.4% in Italy and Switzerland, respectively. This discrepancy is especially significant when considering the timing of the surveys in this study. In the United States, the survey ran just three weeks into stay-at-home orders, whereas Italy had already been experiencing lockdowns for six weeks and Switzerland for five when people responded to the questionnaire. In other words, in about half the time, Americans in the study were three times as likely to have been laid off—an extreme discrepancy.

These divergences are a clear reflection of differences in national policies. The Italian government put in place various provisions to protect its workforce during the pandemic such as special funds to supplement earnings and social security contribution exemptions for companies.[17] In Switzerland, instead of laying people off, employers often shifted employees to so-called reduced working hours for those who could not perform their work from home resulting in temporary reduced pay rather than being let go.[18] While the United States eventually implemented assistance, it took longer and worker protections are generally weaker in that country to begin with, a problem compounded by its system of employer-dependent health insurance, of particular concern during a health crisis.

At the time of the April surveys, it was not yet evident just how long working from home would become many people's reality, and so the study did not ask about related experiences. The May survey in the United States did ask, however. Among the 59% of respondents who were in the labor force, 44% said that such arrangements were not possible with the type of work they do and 6% reported that their employer did not allow it. This left half of those in the labor force working from home because of COVID-19. However, for a quarter of this group this was not a new experience.

Both higher education and higher income correlate with the opportunity to do one's work from home. The most likely jobs to allow such arrangements are desk jobs, which are often associated with higher socio-economic status. This finding shows the unequal circumstances that the pandemic thrust upon people of different means, with those from less privileged backgrounds more likely to need to continue working outside of the home and thus unable to shelter-in-place to protect their health.

TRUST IN INSTITUTIONS CONCERNING THE PANDEMIC

Much of how the pandemic was being handled depended on institutional actors. Although these are not the focus of this book, having a sense of how much people trusted institutions and their representatives to handle the pandemic can be illuminating. Federal as well as local governments were making policies about lockdowns. Medical systems were bursting at the seams trying to handle the onslaught of cases. The media were constantly updating people about goings on both locally and globally. How convinced were people that such institutions were to be trusted?

The survey administered in the United States and Switzerland (but not the one in Italy) asked people to rate their level of confidence in the medical system addressing the pandemic. Two-thirds of American respondents and 71% in Switzerland reported a high level of confidence. Only a small minority (4%) in both countries said they had hardly any confidence (the rest landed in the middle at "only some confidence"). Asked then to what extent they believed medical researchers and doctors understood the spread and health impacts of COVID-19, 57% of Americans said very well (the top rating of five) with just under half of Swiss respondents saying the same. In both countries, more than three-quarters of respondents picked the top two ratings.

It is noteworthy that in the United States, Black participants rate their confidence in both medical institutions and those working in that field lower than Whites. There is a long history in the United States of African Americans' fraught relationship with the medical system (e.g., their mistreatment in the Tuskegee experiment, the infamous syphilis study), which may explain some of these findings.[19] As Laura Specker Sullivan writes in her investigation of trust and race in American medicine: "For those who have faced exploitation and discrimination at the hands of physicians, the medical profession, and medical institutions, trust is a tall order and, in many cases, would be naive."[20] This is very much reflected in the data here even when controlling for other factors in the analyses such as age, education, income, and rural residence. Asian Americans expressed even lower confidence in the medical system addressing COVID-19.

After the medical system, Americans trust local and state government most followed by the federal government in addressing the pandemic. Business leaders and religious leaders were ranked lower. In Switzerland, the federal government enjoys somewhat more trust than local government with both trusted more than in the United States. The Swiss have less confidence than Americans in business and religious leaders, however.

When it comes to believing that various professional groups understand the spread and health impacts of the novel coronavirus (as opposed to trusting them to address the pandemic, which are the results discussed above), Americans were more optimistic than the Swiss. They gave high marks to medical researchers and doctors followed by local and state-level politicians, federal-level politicians, journalists, business leaders, with religious leaders last. The order is similar in Switzerland, except for the federal government coming in ahead of local governments. Also, the average ratings for journalists, business leaders, and especially religious leaders were lower than in the United States. Interestingly, disabled people in the two countries differ in their trust of the medical system and doctors with respect to the novel coronavirus situation. Disabled Americans reported trusting the professionals running the medical system more to handle the pandemic than other authorities, although there is no difference in their belief of doctors' understanding the health impacts of COVID-19. Among the Swiss, disabled people are considerably less trusting of both the medical system and of doctors.

KNOWLEDGE ABOUT COVID-19

There are few instances where being knowledgeable has as immediate a benefit as during a rapidly developing pandemic. Understanding what to do to avoid infection and what to do when infection is suspected became a crucial public health matter. The media was filled with discussions and recommendations, but these were not always consistent and helpful. Chapter 4 delves into where people got their information about the virus and how this linked to their knowledge about it. The goal here is to describe whether people understood how to remain safe and what to do when infected in order to give a general sense of people's virus-related know-how.

The survey asked some multiple-choice questions for which people had to pick the correct response from four options. (The appendix lists all the answer options provided with their corresponding popularity by respondents.) None of these were trick questions, and all questions concerned guidelines that the World Health Organization was providing at the time and had made available on its website.[21] While it may be that some of these avoidance strategies have since been contested, at the time of the survey, these were believed to be the safest approaches by experts.

The questions asked widely discussed topics about the virus including what someone should do if they had come into contact with an infected person (correct answer: self-quarantine by staying at home as a precaution); the common symptoms of COVID-19 (fever and dry cough); how long it takes between catching the virus and beginning to have symptoms (up to two weeks); who is most at risk of serious health consequences from the virus (older people with certain preexisting medical conditions); and what can be said about those who have been tested positive, but are in good health (they are contagious regardless of whether they show symptoms). Figure 1.6 shows what portion of respondents in each country answered each of these questions correctly.

Viewed from one perspective, the majority knew the correct answer for each individual item, which is encouraging. However, given the deadly nature of the fast-spreading virus, where lack of compliance with safety measures by even just a few could have considerable consequences, it is disconcerting just how many people did not appreciate the need for self-quarantining after coming in contact with an infected person. This was the case with 20% of people in Italy and even more in the United States and Switzerland, at 27% and 28%, respectively. A considerable number also did not appreciate that not showing symptoms did not mean they were not contagious with almost one-fifth in both the United States and Switzerland holding this belief (those in Italy understood this better). The fact that Italians did better on these questions may be a result of it being the first of these countries with severe consequences of COVID-19, where deaths skyrocketed in early March and lockdowns went into effect sooner.

The lack of knowledge depicted here shows why government measures about lockdowns were necessary. Without them, spread of the virus likely would have been much worse given that even after considerable public communication, plenty of people did not appreciate just how contagious the virus was and how much at risk they and others would be if those infected continued to socialize.

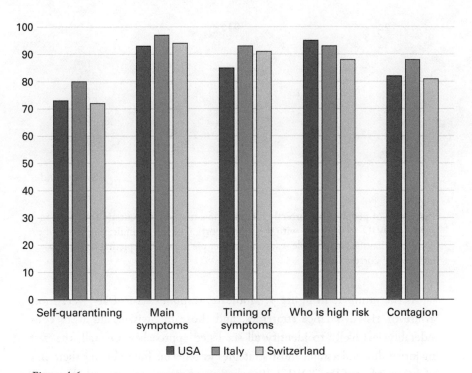

Figure 1.6
Percentage of respondents who chose the correct responses to multiple-choice questions about COVID-19 in the three countries.

The survey probed people's knowledge about avoidance strategies through a different question type as well, this time asking them to check off the various ways in which people could reduce the risk of being infected from a list of ways presented on the questionnaire. The correct strategies were interspersed with incorrect options discussed in the next section about misinformation. In essence, these were true-or-false questions with respondents having a 50% chance of getting any one item right.

The risk-minimizing strategies included washing hands with soap (number correct by country: US: 95%; IT: 92%; CH: 95%); keeping a distance of 6 feet (2 meters) from other people (US: 91%; IT: 93%; CH: 94%); avoiding handshakes (US: 89%; IT: 93%; CH: 93%); not touching one's eyes, nose, and mouth with one's hands (US: 88%; IT: 87%; CH: 95%); cleaning and disinfecting frequently touched surfaces (US: 86%; IT: 78%; CH: 90%); and not leaving the home (US: 83%; IT: 80%; CH: 90%).

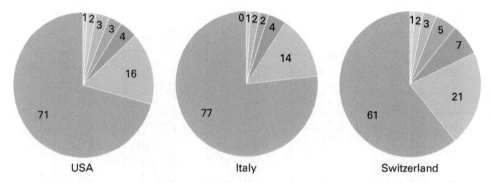

Figure 1.7
Percentage of people who correctly identified in a true-or-false question the ways to reduce COVID-19 infection with the largest portion of the pie indicating those who correctly identified all possible ways and smaller pie segments depicting decreasing numbers of correct answers.

As the numbers in figure 1.7 indicate, respondents in Italy were most aware of risk-avoidance strategies, and those in Switzerland were considerably less likely to identify all six listed approaches. Overall, the vast majority did understand most strategies as relevant for reducing their risk of being infected by COVID-19, which is encouraging. Nonetheless, it is notable that in all cases there are also people who could only identify a few, which is disconcerting given that these were relatively uncontroversial health recommendations already in the early days of the pandemic. Note, for example, that the questions did not ask about masking as, at the time, there was no widespread consensus about that particular approach and it quickly became a political lightning rod. The survey purposefully avoided asking about strategies whose effectiveness was not universally agreed upon at the time of data collection and avoided approaches fraught with political connotations.

From these two knowledge measures—the five multiple-choice questions and the six true-false items—I created a COVID-19 knowledge score with possible values ranging from zero to eleven. This is the score I refer to throughout the book when I talk about people's knowledge related to the pandemic. The average knowledge score was 9.6 among American respondents, 10.1 among Italian participants, and 9.5 among Swiss respondents. As figure 1.8 shows, a substantial portion of the population answered all

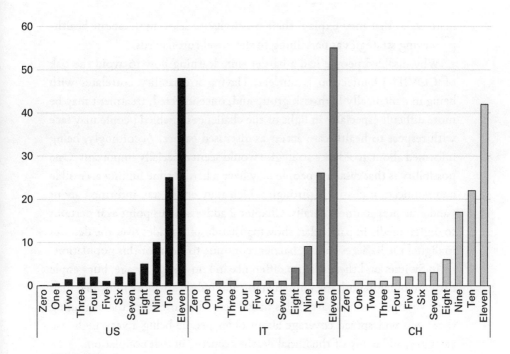

Figure 1.8
Distribution of how many COVID-19 knowledge items people answered correctly in the three countries by percentage.

eleven knowledge items correctly (US: 48%; IT: 55%; CH: 42%) with Italian study participants doing best and Swiss participants doing worst.

There are distinct patterns in who was more or less informed across the three countries. Chapter 2 shows how digital context associated with knowledge, chapter 3 looks at how social media engagement about the pandemic mattered to understanding the virus, and chapter 4 considers how information sources related to being informed. This section reports only on findings based on sociodemographic factors with the addition of political orientation given how quickly aspects of the pandemic become politicized in various countries. When considering age, gender, education, income, rural residence, and disabilities status (plus race/ethnicity in the case of the United States) simultaneously, in all three countries, women were more knowledgeable and disabled people were less. Why this was the case is hard to say, but traditionally women spend more time seeking information about health matters, so that may explain their proclivity to learn more about health-preserving strategies as pertaining to the novel coronavirus.

Why disabled people had a harder time learning how to avoid the risk of COVID-19 infection is unclear. Having a disability correlates with being in a medically high-risk group and, once infected, treatment may be more difficult especially in light of the challenges disabled people may face with respect to health care access as discussed earlier. Accordingly, being informed about avoidance strategies would seem especially important. One possibility is that disabled people may have a harder time finding accessible communication channels through which they could stay informed about pandemic preparedness details. Chapter 2 addresses this point as it pertains to digital media in particular, showing that despite strides over the decades in digital media accessibility, barriers continue to exist for this population.

In Switzerland there was no difference in knowledge by age, but people fifty and older in both the United States and Italy were more informed than their younger counterparts. This makes sense because from the beginning there was widespread coverage about older people being a very high-risk category, and many of the initial deaths occurred in that population.

In the United States, higher income is associated with more knowledge, whereas in Switzerland, higher education links to being more informed (neither relates to knowledge in Italy). One would expect education to be linked given that health literacy is generally related to higher literacy levels. That said, people participating in a written survey such as this study would likely have basic literacy levels, so those on the lowest spectrum of reading ability and comprehension are likely not present in the data set.[22] The relevance of income for pandemic knowledge in the United States may reflect that those in the most precarious financial situations in a country where social safety nets are weak may have been too busy focusing on how to secure continued income to the detriment of following COVID-19 recommendations closely. This is in line with earlier discussions in this chapter about how the three countries differed in dealing with assisting those whose jobs could not simply shift to being done at home.

Differences in pandemic knowledge by race and ethnicity in the United States reveal African Americans and people of Hispanic origin having less knowledge about how to stay safe. Indeed, the difference in knowledge about the pandemic between Blacks and Whites is especially pronounced for those African Americans who said they hardly have any confidence in the medical system addressing the pandemic as was discussed earlier in this chapter about institutional trust. Research has also found similar distrust among Mexican-Hispanics in the United States, suggesting that such mistrust extends to other racial and ethnic minority groups and thus may explain differences in knowledge there as well.[23]

Even a health crisis brought on by a killer virus that does not care about the political orientation of those infected could not withstand being politicized rather quickly. From early on, politicians and their supporters disagreed on numerous important points from the pandemic's severity to what precautions and government interventions people deemed preferable and advisable.[24] Given that politics mattered to how people approached the pandemic and guidelines for staying safe, it is important to consider how knowledge about it may have varied by political orientation among study participants. Although the political systems in the United States, Italy, and Switzerland differ, they all have people with varying levels of left- and right-leaning ideologies. (See "Survey Questions Used in the Analyses" in the appendix for how this was measured in the study.)

Among US respondents, there is no relationship between political orientation and COVID-19 knowledge whether taking sociodemographics into account or not. In Italy, the two are not correlated either, although when controlling for sociodemographics, the significance of political leaning is marginally significant so that those on the right are somewhat less knowledgeable. The situation is different in Switzerland. There, those on the right are slightly less knowledgeable than those on the left, both in bivariate and more advanced analyses. Including political orientation in the models does not change the findings about other sociodemographics discussed earlier in this section.

In a pandemic where a deadly virus spreads quickly from one person to the next, any one individual's knowledge about how to avoid getting infected and how to avoid passing on the infection to others is directly linked to the population's health at a larger scale. Given these network effects of COVID-19, achieving an informed citizenry is in everybody's interest. Understanding facts about the virus is crucial, but it is also important to avoid misconceptions about it because misinformation can have severe consequences too. The next section looks at who was more or less likely to believe incorrect information that was circulating about the pandemic.

MISINFORMATION BELIEFS ABOUT COVID-19

Part of the reason it was initially difficult to be informed about the pandemic was the amount of confusion surrounding the issues and the considerable outright misinformation that circulated about it from the start.[25] How widespread were certain misbeliefs? In this section, the main aim is to present the baseline distribution of different kinds of misinformation beliefs in the three countries to give a general sense of their prevalence. Like in the previous section, many of these were misconceptions addressed on the World Health Organization's website at the time.[26]

Amid the confusion and cacophony, there was variation in what people believed would help avoid COVID-19 infection (the list here replicates the exact wording of the survey): taking vitamin C (US: 35%; CH: 22%; IT:13% believed this); avoiding buying products made in China (US: 20%; IT: 3%; CH: 9%); drinking hot fluids (US: 20%; IT: 5%; CH: 9%); avoiding receiving packages from the postal service (US: 16%; IT: 4%; CH: 8%); taking hot baths (US: 16%; IT: 4%; CH: 4%); avoiding physical contact with pets and other animals (US: 16%; IT: 3%; CH: 9%); avoiding taking anti-inflammatory drugs (US: 13%; IT: 14%; CH: 22%); frequently rinsing nose with saline (salty water) (US: 12%; IT: 8%; CH: 6%); eating freshly boiled garlic (US: 6%; IT: 2%; CH: 5%); avoiding consumption of meat products (US: 5%; IT: 2%; CH: 2%); and avoiding consumption of dairy products (US: 4%; IT: 1%, CH: 2%). In most cases, Americans were considerably more likely to hold misconceptions than Italians and the Swiss.

Why might believing any of the above be of concern? After all, there is little risk in taking vitamin C, drinking hot fluids, or taking a hot bath. Belief in these approaches as a way to reduce risk of COVID-19 is significant because it may result in a false sense of safety. If people think taking a hot bath can help them prevent infection or help in the case of exposure or getting sick from COVID-19, they may be less inclined to abide by guidelines for physical distancing and follow other preventive measures.

Other misconceptions can have negative consequences at a broader level. Believing that buying products made in China is a problem may correlate with targeted animosity. Indeed, while the rise in hate crimes against Asian Americans and Pacific Islanders in the United States came at a time when racial tensions were more generally prevalent, including an increase in hate crimes in 2019, there were plenty of incidents targeting those groups linked anecdotally to COVID-19 misconceptions specifically.[27] The fact that 20% of American respondents believed that by avoiding buying products made in China they could reduce the risk of infection is considerable and is in stark contrast with the 3% in Italy and 9% in Switzerland who held this belief.

Figure 1.9 shows the relative proportion of people believing various amounts of misinformation in the three countries with the leftmost number signaling the percentage of those who did not hold any misconceptions the survey asked about. Almost two-thirds of Italians (64%) did not believe any of the false claims, which is encouraging. More troublesome is that this number is under half in Switzerland and the United States, with just over a third in the latter. Indeed, about 14% in the United States believed at least four of the misconceptions. Similar to knowledge about the pandemic, Italian respondents were the most informed about "fake news" as well, meaning that they did not believe it. This may again be due to the survey having been run at a slightly later point in their pandemic timeline

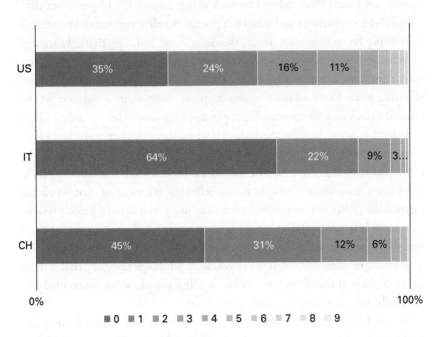

Figure 1.9
Percentage of people who held different numbers of misbeliefs about the pandemic in the three countries. The zero category on the left of each bar indicates those who indicated no misconceptions about COVID-19.

than in the case of the other two countries, and the extreme relative severity of the pandemic in that country in those early weeks.

On the whole, the survey measured belief in eleven misconceptions, so respondents' "fake news" scores range from 0 to 11. The average among American participants was 1.6, among Italians 0.6, and among the Swiss 1.0.

It is important to note that knowledge of and misconceptions about the pandemic, while correlated negatively, are by no means two sides of the same coin. Yes, on average, those who know more are less likely to hold misinformation beliefs about the virus, but there is no obvious overlap. Statistically speaking, in the case of the United States and Switzerland, one explains only about 7% of the variation in the other. In the case of Italy, this is higher at 15% of the variance explained, meaning that the two were more correlated there. By way of comparison, looking at the relationship of education and income among study participants, in the United States, education explains 17% of the variation in income. Put another way, education and income are much more related to each other among US respondents than COVID-19 knowledge and misconceptions. So although more knowledge does imply fewer misconceptions, this association is not particularly strong.

What explains the propensity to believe misinformation? The two consistent findings across countries is that younger people and disabled people were more likely to have misconceptions about virus avoidance. In the United States and Switzerland, people age fifty and older were less likely to believe misinformation; in Italy, those in their thirties were less likely than those in their twenties. Although research has found older adults to be more prone to sharing misinformation on services like Facebook (which, to be sure, does not necessarily mean believing it), most of that work has focused on political content and thus may not generalize to health-related matters.[28] It is noteworthy that with content especially important for their well-being, older adults are less inclined to believe incorrect information than younger ones; indeed, it is the youngest adult age category that is most likely to have misconceptions. Why disabled people were more likely to have held misinformation beliefs is not clear.

In the United States and Switzerland, women were considerably less likely to hold misconceptions about the pandemic; there are no differences by gender in Italy. No differences exist by educational background or income level in any of the three countries, which shows just how indiscriminately misinformation spread across populations. Political orientation is where differences arise in two of the three countries. At first, when only looking at the relationship of political slant and "fake news" beliefs, there is no connection in any of the countries. However, when controlling for sociodemographic background, in both the United States and Switzerland, those on the right held more misconceptions than those on the left. As time passed, political polarization became more evident in some cases. In the May survey administered in the United States, we asked a true-false question about whether "the drug hydroxychloroquine can help against COVID-19," an idea that had been popularized by President Trump, but challenged by medical experts.[29] The response to this question very much varied by political orientation with those on the right significantly more likely (57%) to claim the statement was true compared to those on the left (29%), a statistically significant difference regardless of whether the analyses held sociodemographics constant.

LIFE DURING UNSETTLED TIMES

It has become somewhat of a cliché to suggest that the events of 2020 were unprecedented, but the reality is that few had experienced the types of disruptions to daily life that were now impacting people of all walks of life. A tiny invisible virus was raging across the globe with very visible effects. Thirty percent of Italian participants, 32% of the Swiss, and 18% of Americans in the study reported knowing someone who had been diagnosed with COVID-19. In the latter case, a month later, 26% reported the same. Two percent each in the United States and Switzerland, and 1% in Italy, had been diagnosed themselves. Needless to say, those severely impacted would not have been filling out a survey and so are missing from the data here. When asked whether anyone they knew had died of the disease, 6% in the United States in April and 8% in May answered in the affirmative. In Switzerland this was the case for 11% of respondents. In Italy, a very considerable 21% knew someone who had died from COVID-19. All this just in the first few weeks of the pandemic. In such cases, the virus was no longer something abstract, portrayed in the media as happening far away; it was real, tangible suffering and danger much closer to home.

As discussed in this chapter, people across the United States, Italy, and Switzerland were experiencing the pandemic in varying ways. Some were worried about the most basic of human needs such as access to food and medications. Others saw no reason to be concerned about their finances or future plans. Some people experienced positive outcomes in their home circumstances while others struggled to avoid tensions and find needed alone time. In the United States more than the other two countries, people lost their jobs while many others continued to leave home for work despite the health risks that entailed. Amid the cacophony of information circulating about the virus, people's understandings of it varied, as did their propensity to have misconceptions about how to stay safe. To keep abreast of updates and connect with others, most people had to rely on digital media both for interpersonal communication and information seeking. The rest of this book explores the vital roles that information and communication technologies played in people's unsettled lives during lockdowns. Chapter 2 starts this off by giving an overview of people's digital contexts in these unusual times.

THE DIGITAL CONTEXT OF LOCKDOWN

As mobility restrictions descended on large parts of the globe in March 2020, people became especially reliant on digital media to address basic day-to-day needs from procuring groceries and medications to being in touch with loved ones, colleagues, and friends. This chapter looks at people's technological contexts during this early lockdown period. As outlined in the introduction, even among Internet users, people's digital circumstances may vary considerably. What types of devices were available for people to use? How do people of different sociodemographic backgrounds differ in their autonomy for accessing the Internet? How do digital skills relate to navigating the special circumstances? To capture some of the technological limitations people may have been experiencing, the chapter also looks at who needed, who received, and who did not receive help with their technology needs.

STRUGGLES WITH SWITCHING TO LIFE ONLINE

For those who live in a household with multiple devices and a reliable, fast Internet connection, it may be hard to imagine what life is like for those who have precarious access. Technical limitations can come from the number and quality of devices that members of a household can use to go online and the quality of their connections. Unstable connectivity can stem from only having one device, only having outdated machines, having limited data plans, or living in an area that is not serviced by high-speed Internet options. While the vast majority of people in Western democracies ostensibly have access to the Internet, as the introduction outlined, considerable variations may exist in people's on-the-ground connectivity especially when this is mostly restricted to what is available in the home.

As popular coverage about the pandemic highlighted, many students and adults alike were left coming up with creative solutions to meet their connectivity needs such as spending time in the parking lots of libraries for gaining access to Wi-Fi necessary to complete their school assignments or work obligations.[1] According to data from the US National Telecommunications and Information Administration (NTIA), millions of school-age children in the United States live in households without Internet access. This amounts to 14% of those aged six to seventeen in 2017, the most recent figures available from the NTIA.[2] These proportions are higher among Hispanic and African American households as well as rural areas.

The RAND Corporation, a nonprofit, nonpartisan policy institute, administered a survey of teachers in May–June 2020 to assess inequalities in students' Internet access. A key finding was the vast difference between high-poverty and low-poverty schools in students' home access to the Internet.[3] While 83% of teachers in low-poverty schools reported that their students had such access, a mere 30% of teachers in high-poverty schools could say the same. This not only hinders the ability of students to do their work, but also the ability of teachers to communicate with parents. Not surprisingly, the study also found that students with home Internet access were more likely to complete their assignments and theirs were also the families with whom teachers could more readily be in touch.

In the early days of US lockdowns, the School Superintendents Association administered a nationwide survey of superintendents at public schools to ask about device and Internet access options among their students.[4] Only about a quarter reported that the bulk of their students were in households where they had the devices and Internet connectivity necessary to do their learning online. At the same time, the vast majority of schools were using their websites (97%) and social media (91%) to communicate information to their constituents with considerably fewer turning to more traditional means of communication such as the local newspaper (36%), local radio (21%), or local television (14%). When asked what barriers they faced in transitioning to fully online education, 81% of superintendents referenced people's lack of Internet access at home. These challenges continued well past initial lockdowns as evidenced by the tweet reprinted in figure 2.1 posted in August 2020 by Monterey County California Supervisor Luis Alejo depicting two school children sitting on the street with their devices using the nearby Taco Bell's free Wi-Fi. Alejo deplored the state for not doing better with offering broadband access to students. The tweet went viral and received considerable national media attention showing just how much it resonated with people.[5] This anecdote and the aforementioned figures from various studies show that a basic digital divide in home Internet

Luis Alejo 🔼
@SupervisorAlejo

2 of our children trying to get WiFi for their classes outside a Taco Bell in East Salinas! We must do better & solve this digital divide once &for all for all California students

CALIFORNIA NEEDS A UNIVERSAL BROADBAND INFRASTRUCTURE BOND FOR OUR STUDENTS
link.medium.com/7lr6Dyo5f9

👤 Gavin Newsom and 7 others

8:38 AM · Aug 26, 2020 · Twitter for iPhone

909 Retweets   **150** Quote Tweets   **1,483** Likes

Figure 2.1
Tweet by Monterey County California Supervisor Luis Alejo about school children using Taco Bell's Wi-Fi to do their schoolwork. Tweet with name reprinted with permission. (Photo by Olga Guzman.)

access continues to exist in the United States, impacting students from different backgrounds at varying rates.

Many districts poured considerable resources into making sure that students from different backgrounds would have adequate Internet access. Chicago, for example, spent tens of millions of dollars on the effort, but still ran into problems.[6] As the nonprofit educational news organization Chalkbeat reported, some parents who received an email about the availability of free high-speed Internet access dismissed it as too good to be true. Another barrier to such access was the quick change in household incomes. While families could only qualify with certain income levels, the system was not immediately able to recognize recent changes in financial circumstances brought on by the pandemic such as parents having just lost their jobs due to COVID-19. This then left some of those who qualified without immediate access to the program. Among those who were able to sign up, connection quality remained an issue, which is a considerable impediment to accessing school when many sessions require live video connections.[7]

In some districts, school officials got creative and worked with private companies to transform school buses into Wi-Fi hotspots.[8] Fortunately, some such initiatives had already been in place (e.g., the SmartBus solution by the Kajeet company) offering Wi-Fi to students en route from home to school and back. Such services, available from California to Texas to Indiana, were now able to pivot to offering a Wi-Fi hotspot while stationary in a specific place. Of course, even this posed its own set of challenges such as offering enough hotspots to avoid too many students congregating in one place and compromising necessary physical distancing requirements.

School-age children were not the only ones impacted by limited Internet resources in the United States. The New America Higher Ed Survey asked college students in August 2020 how much of a challenge having access to stable high-speed Internet connection had been as learning had shifted online.[9] More than half (57%) reported this to be a challenge, with 19% reporting it as a *big* challenge. This was especially pronounced among Latinx college students who were more likely (24%) to indicate such constraints.

Even having a high-quality camera and microphone on a computer cannot be taken for granted. Indeed, half of college students noted that as an issue as well. Two in five reported having had to make unanticipated purchases to allow continuation of their education online, in most cases for a computer or laptop, a headphone or microphone, or a printer.[10] Among students with disabilities, many were left without accessible materials such as those with sight impairments being given reading materials not compatible with their screen readers.[11] Similarly, lectures may not have had the needed captioning or sign-language interpretation for those with hearing impairments.

As mentioned, one way people were meeting their basic connectivity needs was by going to parking lots of public libraries to access Wi-Fi.[12] Those without cars were advised that they could sit within a certain distance of the establishment for access.[13] Gina Millsap, the chief executive of the Topeka & Shawnee County Public Library in Kansas, was quoted in the *New York Times* as saying: "Broadband is like water and electricity now, and yet it's still being treated like a luxury."[14] That quote is reminiscent of a comment made by then Chair of the Federal Communications Commission Michael Powell in a speech in 2001 when he contested the idea that there is a problematic digital divide: "I think there is a Mercedes divide, I would like to have one, but I can't afford one."[15] Millsap's comment reflects that little progress has been made in two decades in shifting the perception of Wi-Fi from being a luxury good to an essential good. The pandemic situation brought the consequences of such policy inaction to the fore.

The United States was not unique in its struggles to move work and school online to ensure ongoing access for people from varying socioeconomic backgrounds amid lockdown measures. As the major Italian daily newspaper *Repubblica* reported, 12% of Italy's students did not have a computer at home, a figure that reached 20% in the poorer southern regions of the country.[16] A considerable 57% reported not having their own machine and thus having to share with other family members. To address connectivity issues, some politicians called on the prime minister to provide free and unlimited Wi-Fi access to all citizens, although this did not materialize.[17] Nonetheless, the country spent close to half a billion Euros supporting distance education, including the purchase of hundreds of thousands of devices to loan to pupils.[18]

A report published by UNICEF also found cause for concern with more than a quarter (27%) of parents surveyed in Italy indicating a lack of sufficient devices to meet the work and educational needs of their families.[19] The same report found that the vast majority of instruction concerned video conferences, which again highlights the importance of quality Internet access in addition to available devices. To fill the void in connectivity, the national public broadcaster, RAI, stepped in to help. Through its "La scuola in tivù" ("School on TV") initiative, RAI started providing multiple daily half-hour classes on various topics including foreign-language learning, philosophy, and informatics.[20] While these classes were available online for those with Internet connection, they were also viewable on TV for those with limited online accessibility. This solution to limited connectivity shows the utility of having a national public broadcaster that can have uniform programming across the nation in the public interest.

Of the three countries that form the basis of the empirical evidence for this book, Switzerland does best on basic Internet connectivity indicators. Proportionally, more of its general population is connected than is the case in Italy and the United States, as noted in the introduction. Indeed, when searching for media coverage of access differences in Switzerland during the pandemic, the piece that came up was an article by the Swiss broadcaster SRF concerning the US situation of parking lot access described earlier rather than reporting on similar local hardships. Coverage in Swiss media about Internet struggles mostly focused on whether the network infrastructure could withstand the amount of traffic that was suddenly required during the day with so many people—adult workers and homeschooled students— simultaneously active online.[21]

Despite its overall better numbers, differences in access options nonetheless exist in Switzerland as well. A study of learning during COVID-19 administered in Austria, Germany, and Switzerland together found that 10% of parents reported not having enough equipment in the home to meet both parents' work needs and students' study needs.[22] More than a fifth (21%) of students noted the need to borrow devices from their parents or siblings to get their school work done. (While these data are for the three countries combined, they all have similar Internet use indicators, so findings about the three are likely a good indicator of the situation in Switzerland.)

Needless to say, having to sit in one's car or on the sidewalk to access Wi-Fi is worlds apart from having access on multiple devices in the comfort of one's home. Having to plan one's online activities to coincide with the school bus hotspot parking near one's location is a far cry from having Internet access in one's bedroom around the clock. This is why the concept of digital inequality is important because it recognizes that even among those who are connected and are regular Internet users, major differences remain in day-to-day online experiences.

Taken together, the combination of study reports and anecdotal evidence presented in the press suggests variation in all of the three countries examined in this book regarding people's ability to pivot to life online. Simply being an Internet user does not automatically mean that one has unfettered, constant, high-quality access. To signify the importance of being able to go online when and where one wants, the literature on digital inequality refers to variations in "autonomy of use"[23] or "next generation Internet users,"[24] which the following section will explore in detail regarding the participants in this study.

AUTONOMY OF USE

A helpful term for thinking about the aspects of digital inequality that give people freedom to use the technology when and where they want to is *autonomy of use*.[25] A user who has the latest devices capable of running the most up-to-date software with high-speed connections and no data limits without others clamoring for access to the same device will have considerably more autonomy for their Internet use than one who does not. Under lockdown, when people are expected to do as much as possible from home, having children on Zoom for school in parallel to parents on Zoom for work may tax the Internet connectivity of a household, making it difficult to tend to one's own responsibilities if the quality of connection is low.

Even more constrained are households where there are not enough devices to go around for simultaneous use. In families with more than one school-age child, it may be difficult for everyone to do their work if access happens on one family computer. In case a device encounters a problem, the owner of multiple gadgets can pivot to using another one; not so if there are not enough machines to go around.[26] Even if there are enough devices, their quality may not be the type to withstand any and all uses. To distinguish between these types of access options, Internet scholars Bill Dutton and Grant Blank came up with the concept of "next generation Internet users," which they define as people who have three or more devices to go online, one of which is mobile.[27] This is in line with the autonomy of use concept as this level of device access grants the type of flexibility (i.e., autonomy) that is beneficial for users.

Communication scholar Amy Gonzales writes of the importance of technology maintenance, referring to the idea that low-income people struggle with unstable access that regularly leads to periods of disconnection.[28] This can be a result of people not having the resources to repair or replace broken devices or it may be caused by periodic loss of connectivity. Such precarious conditions of connectivity are stressful under the best of times. They are likely to be even more disconcerting at a time when people cannot easily turn to public access points such as libraries, schools, and community centers to supplement their limited personal technology resources.

At the other end of the connectivity spectrum are households with multiple devices for each inhabitant, from smartphones to smart TVs, from laptops to gaming devices, from voice assistants to wearable devices, from large curvy monitors to document cameras, from tablets to state-of-the-art printers—all with the potential to be on reliable high-speed Internet connections. While such a fully wired household is not common, it exists and shows the vast variation in people's digital contexts. Such an abundance of cutting-edge devices provides for an entirely different experience, especially at a time when most activities are constrained to the home. In between these two poles of autonomy of use are households with a mix of old and new devices, some of which are shared, with reasonably stable, but not necessarily unlimited data access.

The general statistics described earlier in this chapter and the anecdotes from the media both point to considerable variation in how people live their day-to-day digital lives and what challenges the pandemic brought for their digital context. To examine varied access during lockdown more systematically, I now turn to the types of devices through which participants in the study for this book accessed the Internet at home. It is worth noting that this is a relatively well-connected group. After all, these are people who were able to participate in a survey on a device. Those under the most precarious and least connected circumstances are unlikely to show up in a study that was administered online given that they are least likely to have the technical conditions to do so. This is important to keep in mind to recognize that the general population likely includes more people at the lower end of the connectivity spectrum than those who participated in this study. If anything, this means that any findings here about digital inequality are likely to underestimate differences across the whole population because they will be based on fewer cases of people in precarious technical contexts.

Table 2.1 shows that by far the most ubiquitous Internet access for people is through their mobile devices in all three countries, followed by computers, tablets, smart TVs, and gaming devices. It is notable that about a fifth of respondents have Internet access on all five of these devices in all three countries. These households constitute the upper echelons of autonomy of use given that with so much redundancy, barring any systemwide failures such as power outages, their Internet access is dependable. In case one device breaks, there are several others they can utilize.

Table 2.1
Percentage of respondents with different types of home Internet access in the
three countries

	United States	Italy	Switzerland
Mobile	90	97	94
Computer	82	94	93
Tablet	54	63	64
Smart TV	52	57	49
Gaming device	37	41	31
All five above	19	23	19
Mobile + Tablet + Computer = High autonomy	44	60	58
Only one	12	4	6
None	0.3	0.1	0.1

Given that smart TVs and gaming devices are unlikely to be the most useful with certain tasks, the bottom part of table 2.1 also shows what portion of each country sample has Internet access on a mobile phone, a tablet, and a computer (recall the definition of the "next generation Internet user" described earlier).[29] These are the people who have three devices with reasonable writing and display capability to interact with online information, and at least two of these allow for locational flexibility. Under half of respondents in the United States and just over half in Italy and Switzerland fall into this category. At the other end of the spectrum are people who have only one mode of home access to the Internet: 12% in the United States, 4% in Italy, and 6% in Switzerland. In each country, there was only a handful of respondents who have no home Internet access.

Are there any differences by demographic and socioeconomic indicators in people's digital contexts? With autonomy of use, it is informative to know who has access to several devices for using the Internet. Consistent across all three countries is that younger people and those with higher household incomes, as well as those with children in the household, were more likely to have all five devices as an option for connectivity. In the United States, the circumstances of those with different education levels also varied: the more educated were considerably more likely to have several access options, as were men in general. These findings are robust to both bivariate and more advanced statistical analyses that hold other factors constant. There was no relationship between education and device access in Italy and Switzerland.

Arguably, it is not necessary to have a smart TV or gaming device to accomplish many online tasks, so it is also helpful to consider who has the three primary connected devices—mobile, tablet, and computer—at their disposal. The story here is similar, although not entirely consistent with the aforementioned findings. Household income continues to be a notable correlate of autonomy of use across the three countries, as is having children in the household. Interestingly, however, age is more complicated now. In the United States and Switzerland, there is no difference among older and younger respondents in this regard, whereas in Italy, those in their sixties and older still have fewer device options. Education remains an important positive correlate of autonomy of use in the United States, and in the case of having access to three devices, it is also significant in Switzerland.

At the opposite end of the spectrum of user autonomy are those who only have Internet access at home through one device. Undeniably this makes for a very different type of access, given that any issue with the device would then significantly hinder access to the outside world. Here, there is only one consistent factor across the three countries, one that is not at all surprising: those with lower household income are more likely to have precarious home Internet access as they only have one device for it. In the United States and Switzerland, age also matters, with older people more likely to be in this situation. In the United States, an additional factor is education: those with the lowest levels are most likely to have the fewest access options. In both Italy and Switzerland, disabled people are more likely than others to have just one point of access to the Internet at home. Taking all of these together, it is clear that those from higher socioeconomic status have the most autonomy for accessing the Internet at home on multiple devices, and those in the least privileged positions have the least autonomy.

While the link between lower income and precarious Internet access is not surprising, it is nonetheless worthy of serious consideration. For people who are at the fully wired end of the spectrum, it is all too easy to forget about the diverse circumstances under which people go online. If public officials, educators, providers of various services, and employers do not appreciate these differences in Internet access, they will not set the right expectations and create the needed circumstances for their constituents, students, and employees to meet their needs. If a teacher assumes that all pupils can hop on video any time of day and can print out assignments they are sent, children who do not have the necessary digital context to do all that will fall behind. If an employer assumes that everyone on their staff can easily download and install programs onto their machines and have various desired computer peripherals at their disposal, an employee who operates under more limited conditions will have a hard time meeting the basic needs of their job.

Having better autonomy of use means being able to go online more frequently.[30] Those who have limited access likely are not on their devices all the time, especially if they are sharing these with others in the household. Figure 2.2 shows the large variation in daily Internet use among those who only have one access device compared to those who have several. In all three countries, the differences are considerable. In the United States and Switzerland, the difference is greater than threefold. In Italy, this divergence is even more pronounced; while only 1% of those who have multiple devices at home with Internet access only go online once a day or less, 13% of those who only have one such device fit this description. This chapter will return to this point of varied autonomy of use as it investigates other digital inequalities regarding people's technological context during lockdown.

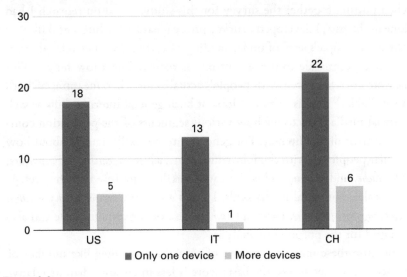

Figure 2.2
Percentage of respondents who only go online once a day or less by autonomy of use (number of devices).

To appreciate just how much digital inequality permeates people's lives, it is important to recognize that already at the most basic level of access, users come to their online experiences under vastly divergent circumstances. In all three countries, those with the least financial resources are in the most constrained situations when it comes to accessing the Internet. While digital media became even more essential than usual during the COVID-19 pandemic, people with the least resources were already starting from behind. They were also probably the least likely to have the needed know-how to expand their home access options.

DIGITAL SKILLS

As the introduction explained, having an awareness and understanding of technologies is an important aspect of using digital media in beneficial ways and is also significant for avoiding problematic uses such as falling for scams and misunderstanding information.[31] Measuring digital skills is not as clear-cut as measuring device availability, however, because there is not something as tangible as number of gadgets to ask about and count. Fortunately, I was able to draw on two decades of studying people's Internet skills (used here synonymously with the terms *digital skills* and *digital literacy*) when putting together the survey for this study. Based on research I had done in the past, I developed a survey proxy measure for Internet skills that asks about people's level of understanding of various Internet-related terms on a five-point scale to use as a proxy of their online know-how.[32] This measure correlates well with people's actual skills when measures of both are available.[33] In this section, I look at both general Internet skills as well as social media skills to see how various segments of the population compare in their digital literacy. For general Internet skills, I asked about how familiar people are with the following terms: *advanced search*, *PDF*, *spyware*, *wiki*, *cache*, and *phishing*. Then I created a skills score based on the average (1–5 scale). For social media skills, I asked about the terms *privacy settings*, *meme*, *tagging*, *followers*, *viral*, and *hashtag*, also on a five-point scale and also averaged for the purposes of analysis.

Because these measures are not of something tangible like number of points of Internet access, the basic score is less interesting than its relative distribution across population groups. So, while I can report that in the United States and Switzerland, the average score was 3.3 on a 1–5 scale compared to Italy's 3.7, this does not tell us much on the face of it. It implies that the Italian respondents were savvier given that 3.7 is a higher average score than 3.3. It is notable, however, that this pattern across countries is not in line with the European Commission's (EC) measure of skills where Italy comes in as one of the least-skilled among members of the European Union (Switzerland is not a member).[34] There are two potential reasons for this divergence. One is that the EC measures skills by asking about digital experiences rather than abilities, so their measure of "skill" is actually a measure of online experiences. Another explanation is that respondents for this study in Italy may be less representative of the average Internet user in that country than is the case for the other two nations surveyed. This is a reminder about the limitation of gathering data online where participants may have originally been recruited to the survey company's respondent pool through various sites and services that favor those who are more frequent and possibly savvier users.[35]

In the context of discussing Internet skills, it is also worth remembering that participants in this study took the survey online and so are likely more digitally savvy than the average Internet user both with respect to general Internet skills as well as social media skills. Had the survey been administered in person (impossible during lockdowns) or on the phone, the mode of data collection may well have captured a more diverse group of users. As noted earlier, this means that any variations found in this sample are likely to be a conservative estimate of the amount of variation across the spectrum of digital inequality in an even more diverse sample.

Looking at the distribution of skills across the sample in each country can be informative for understanding whether there are many people who are very skilled and just a few who are not skilled at all, whether there is more of a normal distribution, or something in between. In figure 2.3, the line centered around the middle represents a normal distribution, which happens when most people are in the middle of the range (around a 3 score on the 1–5 scale), while fewer are at the ends of the distribution, either very unskilled (1) or very skilled (5).

In reality, the distribution is somewhat left-skewed in all three countries, some more than others. The dashed line represents the United States, where there are fewer people at the low end of the distribution and a similar number of high-skilled users as those in the middle of the pack. Italy (gray line) has the most skewed distribution with relatively more people with higher

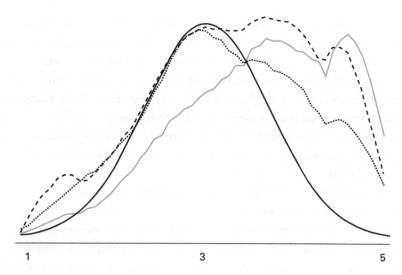

Figure 2.3
Internet skills distribution in the three countries (U.S. = dashed; Italy = gray; Switzerland = dotted) with a normal distribution curve in black as reference.

scores. Switzerland (dotted line) is the closest to a normal distribution, because the highest peak is in the middle while the line trails off on both the low and high end of the range (although trails off less so at the high end). The main takeaway is that although lower-skilled users are not as common as others, there is certainly variation in all three countries in how skilled people are with the Internet. Social media skills (not pictured) are similarly left-skewed in both the United States and Italy, whereas the distribution is much closer to normal in Switzerland.

Some of the most widely held assumptions about digital literacy concern its relationship to age.[36] One such assumption is that young people are automatically savvy with technologies given that they grew up with them while older people necessarily lack digital prowess.[37] These two age-related suppositions come with various implications. One is that if young people are universally digitally savvy they do not require training and support. The other is that young people are necessarily more skilled than older people and thus the former cannot learn from the latter group, and the latter group necessarily trails behind and is dependent on younger counterparts for assistance. Considerable scholarly work has debunked these myths, but they persist in the popular imagination and rhetoric.[38] This is unfortunate because it shortchanges people of all ages. If young people are assumed to be generally savvy with technologies then schools and employers will not think to offer training and support even though young adults may well need such support.[39] If older people are assumed to be generally clueless with technologies, then opportunities will be lost in how they can support each other and even younger cohorts in gaining digital skills.[40]

What does the relationship between digital literacy and age look like in the people surveyed for this book? Figure 2.4 shows the average skill score by age decade (the lowest group includes those age eighteen and nineteen in addition to those in their twenties and the highest group includes everyone seventy and older). While it is indeed the case that those in their seventies and above score the lowest, there is no obvious uniform declining trend from the youngest group to older groups. In fact, when comparing those in their forties in the United States (leftmost bars) to those in their twenties and younger, the older group reports higher skills. In Italy and Switzerland, there is no statistically significant difference across these groups, so widespread assumptions do not hold there either. This debunks both the myth that younger people

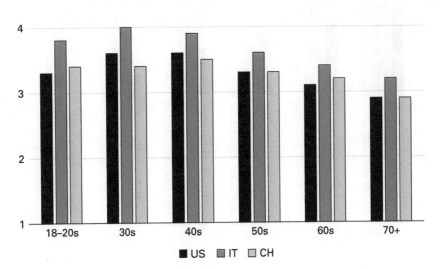

Figure 2.4
Average general Internet skills by age groups across the three countries.

are universally savvy and that older people are necessarily less skilled than younger adults. The assumption holds for the oldest of the old, but only them.

In the area of social media skills, young adults indeed do better than older ones. Figure 2.5 presents these details, clearly showing a downward trend from younger to older groups. In the United States and Italy, there is no difference under age forty; starting at forty, social media skills decline. In Switzerland, it is gradual starting with the youngest adult group.

When it comes to differences by education and income, the story is entirely consistent across the three countries. Those with higher education and higher income exhibit both higher general Internet skills as well as higher social media skills. This is not only true in bivariate analyses, but also when we hold a host of other factors constant such as age, gender, and rural residence. It even holds when we account for whether someone only has one access device at home (as a measure of autonomy of use) and whether they only go online once a day (accounting for use frequency). In the US context, once we take into account having home Internet access on all three general devices (mobile, tablet, computer), then income is no

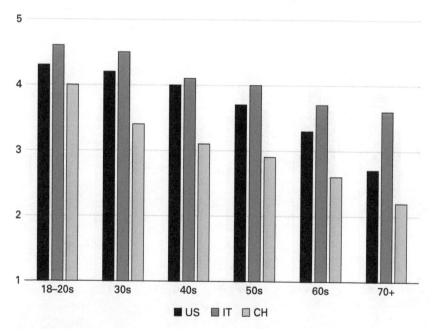

Figure 2.5
Average social media skills by age decades across the three countries.

longer related to skills. This shows that in the United States, higher income matters for skills by way of the technological infrastructure that people have at their disposal at home for going online. In Italy and Switzerland, it goes beyond this financial issue since people with more financial means remain more skilled even when taking autonomy of use into account. In the United States, those who live in rural areas also have lower skills.

Disabled people have for a long time been less connected and shown lower-level digital skills than their nondisabled counterparts for reasons often relating to the inaccessibility of various devices and services. Analyzing data from the US Federal Communications Commission collected in 2009 using a similar Internet skills question as in this book, my long-term collaborator on related studies, Kerry Dobransky, and I found that disabled people had considerably lower Internet skills than others.[41] It is thus especially noteworthy to find that a decade later, this difference no longer seems to exist. Prior work using data from 2005 comparing the Internet skills of people with and without disabilities in other countries also found variations (Italy was included in that study, the United States and Switzerland were not).[42] Across all three countries included in this book, disabled people have similar skills to people without disabilities in 2020. This may reflect their considerable advocacy in recent years for more accessible technologies and workarounds to the limitations of what tends to be available (more on this below). It is encouraging to know that both regarding general Internet skills and social media skills, disabled people are on par with others.

A complicated variable when it comes to skills is gender. Women tend to score lower on the general Internet skills measure yet their actual skills are not necessarily lower than those of men.[43] This is a limitation of the measurement. It is not unique to assessing Internet skills, however. As sociologist Shelley Correll has noted, "cultural beliefs about gender are argued to bias individuals' perceptions of their competence at various career-related tasks, *controlling for actual ability.*"[44] The skill measure used here is still the least biased of other options such as simply asking people what they think their skills are, which is likely to suffer even more from cultural beliefs about divergent gender competencies. Whether a clear reflection of actual skills or not, research has shown that women's assessment of how familiar they are with the Internet relates to what they do online so it remains a valuable measure.[45]

Women in all three countries scored lower than men on the general Internet skills measure. This did not hold for social media skills, however, where women score lower only in the Switzerland sample. The gender differences persist even after taking lots of other factors into account such as age, education, income, rural residence, and disability, as well as autonomy and frequency of use. Overall, what these analyses reveal is that large variation exists in people's level of understanding of the Internet and social media. The rest of the book will explore how this matters for what people do online.

TECHNOLOGY SUPPORT NEEDS

Beyond measuring skills directly, another way to assess whether people are able to take advantage of technologies is to look at the extent to which they need related support. The questions in this section were only on the second survey administered in May 2020, and so are restricted to US respondents per the methodological details discussed in the introduction. The survey included questions about technology needs in two ways, first in a more general manner, and then honing in on specific technical tasks with which people may have needed assistance. The two overlapped although they did not correlate entirely, which underscores the importance of asking about experiences in different ways.

The first question was worded as follows: "During the Coronavirus pandemic, some people have needed help with some activities. Have you needed help with each of the following? For each, indicate whether you needed help and if yes, whether you received the help you needed." Then a list of domains followed, most having nothing to do with technology and thus are not discussed here. (These included needing help such as accessing masks and getting groceries.) The item about technology was "how to do something on smartphone, tablet or computer." The following paragraphs draw on the responses to this latter item.

Interestingly, adults under fifty were more likely to report having needed technology help by the second month of the pandemic than their older counterparts. As figure 2.6 shows, just under a quarter of those in their twenties and thirties answered in the affirmative (23% and 22%, respectively) compared to less than 10% of those fifty and over. It may be that with both social and work life now largely dependent on home access, younger people had to pivot more of their lives to full digital mode than older people. These findings challenge assumptions that younger adults

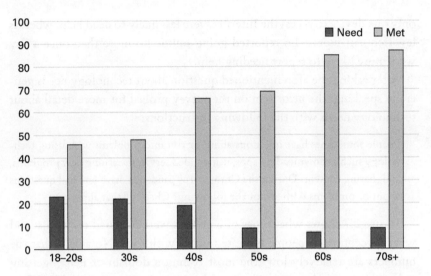

Figure 2.6
Percentage of people who needed help with technology (dark bar) and percent of just these people needing help who had those needs met (light bar) by age group.

do not have technology needs. Also notable is that young adults were less likely to have had these needs met as evidenced by the lighter bars in figure 2.6. These lighter bars represent who among just those people who needed help (dark bars), rather than the full sample, had their needs met. While under half of those in their twenties and thirties were able to get needed support, more than 85% of those in their sixties and over could. Again, this challenges the widely held assumptions that young people can easily address their technical needs and questions.

Perhaps unexpected is that those with higher levels of education were also more likely to need help. It may be that such people were more likely to be in occupations that required pivoting to online availability and depending on technological solutions for doing their jobs. Those with college degrees did not differ from those with lower levels of education in having these needs met, however. Regarding disability status, disabled people were more likely to need support and less likely to have such needs met than those without disabilities, exacerbating disadvantages such populations already face. Women were less likely to need technology help. Level of Internet skills did not relate to needing assistance with how to do something on their smartphone, tablet, or computer during the pandemic. People with only one device to access the Internet were less likely to need help, whereas less frequent users who reported going online no more than once a day were more likely to report needing help.

After asking the aforementioned question about technology needs generally speaking, the next item on the survey probed for more detail about technology needs with the following instructions:

> People sometimes have questions about or run into problems with using technology such as downloading apps, using video services, learning new programs, and sharing content. During the Coronavirus pandemic, have you had questions about or problems with any of the following? Check *all* that apply.

Figure 2.7 shows what types of activities were included on the list and what percentage of people indicated issues with each. Notable is that the numbers are relatively low (the most common domain of help—getting video chat to work—was only listed by 11% of respondents) although it is worth remembering that the question was asked within two months of lockdowns, and therefore the responses reflect initial circumstances of the pandemic not what needs may have arisen in the long term. On the whole, 29% of people reported running into problems or having questions about at least one of the activities listed. Of those who needed help with the listed actions, 42% had just one issue, a quarter needed help with two, 19% with three, and declining numbers listing even more.

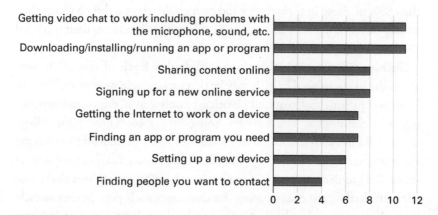

Figure 2.7
Percentage of people who had questions about or had run into problems with various technology-related activities during the first six weeks of the Coronavirus pandemic.

A separate question asked how often respondents were able to find a solution to their questions about and problems with technology during the pandemic. The answer options ranged from "none of the time" to "all of the time," with three other options in between. Of the full sample, just over a third (34%) answered that they had always been able to find a solution; 8% were at the other end of the spectrum, reporting that in no cases did they figure out the answer to their questions.

Similar to the earlier way of inquiring about technology needs more generally, younger people were more likely to have run into problems with the list of specific activities, yet again confirming that they do not necessarily know everything about technologies that may be relevant for them. In terms of whose needs were met, there were no differences by age in this case. Those with higher education had more questions than others reflecting their higher likelihood of needing technology assistance already noted earlier. It may be that people with college degrees were more likely to be in jobs where they were used to getting technology needs addressed by the IT departments of their workplaces even before questions arose, which would have been less readily the case now from afar amid unprecedent circumstances pivoting several people in organizations to working from home simultaneously. Indeed, being in the labor force is positively correlated with having needed help with technology, although it does not explain the differences by education (even when the statistical model accounts for working, those more educated were still more likely to need help). Those more educated were also more likely to find solutions to their needs, highlighting another aspect of digital inequality. Disabled people were more likely to run into the listed issues, but no less likely to find solutions to them (although not more likely either).

Interestingly, users with higher general Internet skills were just as likely to run into problems. This may be because they were trying out more things than those with less know-how and consequently with a higher likelihood of encountering challenges. That said, they were much more likely to find solutions to issues they faced. Those with higher social media skills were more likely to encounter issues and also more likely to report finding answers. On the whole, those more privileged were more likely to report finding solutions to their technology needs than their less-resourced counterparts, signaling that digital inequality extends to support needs.

DISABLED PEOPLE AND TECHNOLOGY ACCESS

As explained in chapter 1, one of the groups of particular interest in the pandemic is disabled people, given their life experiences coupled with the potential of information and communication technologies to assist them. As mentioned earlier, issues of technological access have long been a challenge for this group. Yet, this is an oft-ignored population when it comes to studies of digital inequality.[46] As the following chapters will show, experiences with Internet use often vary by disability status.

Although in the early years of the Internet's mass diffusion, disabled people were much less likely to be online, this has changed considerably over the last two decades.[47] My collaborator Kerry Dobransky and I first documented this in a paper published in 2006, where we analyzed data collected in 2003 by the US Bureau of Labor Statistics and the US Census Bureau. We found that people with all types of disabilities were considerably less likely to use a computer in the home and to live in a household with Internet access. Even among those who had a computer at home that they used and who had Internet access in the home, disabled people were still less likely to use the Internet. When looking at Internet use anywhere, we also found differences by disability status. We surmised that technical accessibility posed a major barrier to this population in making use of digital technologies. Studies covering data from various countries in Europe have found similar patterns.[48] Even when accounting for the socioeconomic differences between disabled people and others, variations in access and use remained.

Most hardware, software, and web content is not created with disabled people in mind, which likely explains the aforementioned findings. These limitations of mainstream options pose a host of challenges for those who are blind or visually impaired or those who cannot use a traditional keyboard and/or mouse.[49] There are certainly mandated and recommended guidelines for making computers and the web accessible, such as the 1998 Section 508 update of the US Rehabilitation Act of 1973, the Americans with Disabilities Act, and the World Wide Web Consortium's Web Content Accessibility Guidelines.[50] However, these are often not followed by device and content creators.[51] Related policies also exist elsewhere: the European Union's Web Accessibility Directive and the European Accessibility Act of 2019 support the need for accessible products and services including computers and smartphones.[52] In Switzerland, the Disability Discrimination

Act was amended in 2002 to include "communication systems," but again, compliance is not a given. This makes it difficult for disabled people to adopt new technologies as they become available.

When it comes to supporting access for disabled people, the focus tends to be on assistive technologies such as screen readers, speech-to-text applications, and other accessibility additions by way of devices and programs.[53] Not only does this delay access to new opportunities, it also tends to come with a considerable price tag. This is especially limiting given that disabled people are often forced into more modest financial means, as is the case among participants of this study among Americans. (The same relationship of disability status to income does not show in the Italy and Switzerland samples.) Facing these challenges of post hoc solutions to accessibility issues, advocates focus on the importance of universal design or universal usability, which encourage accessible design from the outset to the greatest extent possible.[54]

It is not, however, only when particular accommodations may be necessary that disabled people are at a disadvantage in digital contexts. Comedy writer Jackie Quinn tweeted about her frustration over a website she encountered while trying to figure out her access to needed medical equipment during the COVID-19 pandemic. As illustrated in figure 2.8, instead of being a helpful resource, the site had a placeholder template literally saying "something, something, something," which is, of course, entirely useless to people trying to inform themselves about their medical equipment options. Not only are experiences like this frustrating for disabled people—as they would be for anyone encountering such a site—they may also discourage extended use if such resources repeatedly prove unhelpful.

A decade after our 2006 paper about the disability divide in Internet use, Kerry Dobransky and I published another article, this one titled "Unrealized potential: Exploring the digital disability divide."[55] Analyzing 2009 data collected by the US Federal Communications Commission, we showed continued disparities by disability status in using the Internet, having broadband at home, their autonomy of use (number of Internet access locations), and their Internet skills. We also found that this group engaged in fewer activities online than their nondisabled counterparts. Significantly, however, we found that disabled people were more likely than others to submit a review about a product or service. The example in figure 2.8 illustrates this well. With countless frustrating online experiences, it is no wonder that disabled people would want to voice their concerns about services, either because those

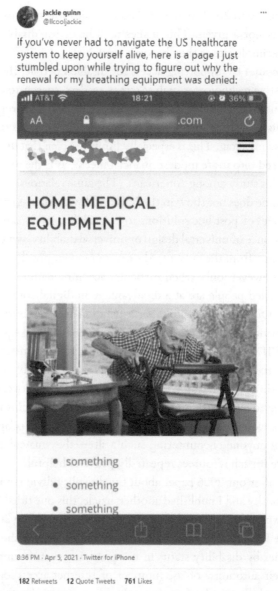

Figure 2.8
Tweet by user @llcooljackie expressing frustration over difficulty with obtaining needed information in light of website with template content ("something, something, something") instead of needed information about home medical equipment. Tweet with name reprinted with permission.

services come up short in meeting disabled people's needs or in case of what are likely to be unusual positive experiences, presumably to help others by spreading the word.

Data collected in Poland in 2013 also showed that disabled people were generally less likely to be online, but those who were online were more active in publishing their own digital content.[56] It is important to recognize that when possible, disabled people take advantage of the opportunity to share their voices online. Such findings show that it is not a lack of interest in online participation that holds this population back from online engagement.

DIGITAL CONTEXT AND UNDERSTANDING THE PANDEMIC

Having established that people vary in their autonomy of use, frequency of use, and digital skills, it is instructive to see whether these then relate to people's understanding of the pandemic by way of the knowledge and misinformation measures described in the previous chapter. All of these results control for sociodemographics in the analyses (age, gender, education, income, rural residence, and disability status, plus race and ethnicity in the United States).

Regarding frequency of use, findings are consistent across the three countries: people who only go online once a day or less know less about the virus while there is no link with misinformation beliefs. Findings are also relatively consistent across the countries for autonomy of use: having only one home device access point links to lower levels of knowledge about the pandemic in all three countries. Having all three access device types (computer, tablet, mobile) is associated with more COVID-19 knowledge in the United States and Switzerland, but not in Italy. Similar to frequency of use, when it comes to misinformation beliefs, level of user autonomy does not matter.

The relationship of digital skills to knowledge and misconceptions about the virus is less straightforward. In the United States, general Internet skills do not relate to knowledge, but social media skills do. In Switzerland, it is the reverse in that general skills matter, but social media skills do not. In Italy, both are positively linked to knowledge. On the whole, then, there is a positive link between digital skills and understanding virus risks and prevention across the three countries.

Holding misconceptions about the virus plays out entirely differently in the three countries. Recall from chapter 1 that when only looking at sociodemographics, older adults were less likely to have misinformation beliefs in all three nations, women were less likely in the United States and Switzerland, and disabled people in all three countries were more likely. Higher general Internet skills link to higher misconceptions in the United States, which is not the expected relationship. There is no connection with social media skills. In Switzerland, there is a curious interplay of age and skills (both general and social-media-related) in that adults sixty and older who are more skilled are less likely to hold misconceptions about the virus, but adults under that age are more likely to believe misinformation when they are more skilled, reminiscent of the findings for the United States. In Italy, general Internet skills show no relationship with COVID-19 misconceptions, while those who have higher social media skills are less likely to hold misinformation beliefs. Overall then, social media skills seem more relevant than general Internet skills for avoiding misconceptions. There has been very little work so far on connecting digital inequality indicators with misinformation beliefs, a domain clearly ripe for more investigations. Chapter 3 looks at how people were communicating on social media to try to unpack some of these findings about social media skills, and chapter 4 then looks at how information sources more generally connected with understanding the pandemic.

DIGITAL INEQUALITY IN THE TECHNOLOGICAL CONTEXTS OF USERS

On the whole, when considering people's access options, autonomy of use, digital skills, and support needs, the story that emerges is one of considerable variation depending on a user's sociodemographic background. Consistent across the three countries is that people of higher socioeconomic status are more likely to have more devices to go online, to be more regular Internet users, to have higher Internet skills, and to have their support needs met.

Challenging widespread assumptions about younger adults' universal savvy with technologies and older adults' lack of know-how in this domain, evidence from all three countries suggests a more complicated picture. Not all young adults are by definition savvy with technologies nor are they able to get all of the support they need. And not all older adults are clueless about digital technologies. While it is certainly the case that those in the oldest age categories continue to have fewer access options, be less frequent users, and have lower-level skills, there is not an obvious linear relationship of age to important digital inequality indicators below age seventy. The online experiences of disabled people are not as some may assume, because they are now online in larger numbers in more active ways and with higher skills than before despite continued barriers that technologies pose. Nonetheless, they continue to have unmet technology needs that may pose barriers to their effective online participation. The following chapters look at how this pertains to communication on social media as well as information seeking about the pandemic.

CONNECTING ON SOCIAL MEDIA ABOUT THE PANDEMIC

Social media like Facebook, Instagram, Twitter, and WhatsApp have become some of the most popular ways for people to connect with others and get information in the twenty-first century. This is perhaps not surprising given everything they allow users to do.[1] Jokes about people oversharing personal details of their lives notwithstanding, such platforms offer many opportunities for their users that can address both personal as well as professional needs and interests. They let people communicate with both lifelong friends and long-lost loved ones while also offering numerous chances to meet new people. They can help explore career opportunities, connect with colleagues, promote a business, or help with personal branding. They can serve as an informational resource on countless topics. Users can exchange ideas and share content on such services one-on-one or in groups.

Social network sites allow people to pursue their various interests from sports to cooking, from religious communities to arts and crafts, from medical advice to disability support, from neighborhood assistance to childcare and eldercare discussions. Users can offer up goods and services for free or for a fee. The platforms can also be used for political purposes through fundraising, organizing, mobilizing, and proselytizing. In sum, for any topic of interest to someone, no matter how niche or obscure, there is a good chance that an online community—or three hundred—exists to bring likeminded people together to communicate about it. And while this is not new to the Internet—indeed, it is one of its oldest functions[2]—social media have made these many types of connecting much more accessible to the vast majority.

Given social media's mass appeal as a type of online platform, this chapter focuses on how people used it to connect about the pandemic. After giving a bit of general context about what makes social media distinctive in the ecosystem of online services, this chapter will illustrate how various such platforms attract people of different sociodemographic backgrounds. This highlights that social network sites should not be thought of as interchangeable and why it is important to gather data about practices on more than one when examining how people incorporate them into their lives. Next, the chapter describes what types of COVID-19-related content people saw and shared on such platforms. The sections that follow then show how such exposure and engagement is associated with social connectedness as well as knowledge and misconceptions about the virus.

Social media are distinct from other modes of communication in that they make relationships among other users visible.[3] If a user interacts with another (except through a private message), other users can see that relation and interaction; they can often also jump in and participate in the same exchange. While social media platforms today allow users to engage in considerably more activities than their variants from even just a decade earlier, a primary aspect of these services remains their social nature: their ability to let people connect and to share content with one another.

Of course, this connectivity has downsides as well.[4] There is no guarantee, for example, that information people see on such platforms is useful or truthful. Interactions on social media will not always be collegial or respectful. Users are likely to encounter perspectives incongruent with their own or comments that offend them, and their tolerance for such challenging content will vary. People may also dislike others' styles of communication, resulting in stressful readings or tense exchanges. There is also the possibility of feeling down about one's circumstances in reaction to seeing what others share, perceiving it as better than one's own situation. Indeed, since many users have a tendency to exaggerate the good and joyful while hiding the unfortunate and uncomfortable in depictions of their everyday lives, feeling inferior in light of what one sees on social media feeds is not unusual, and research has shown it can be downright damaging.[5] Although powerful tools for sharing and reflecting the complex realities of life, social media also have the potential to distort those realities and foment tensions.

Given their broad coverage and reach, it is not surprising that people turned to social media during the pandemic to share and seek content as well as connect with others. This chapter goes into detail about how people used such platforms during lockdown for COVID-19-related content and discussions in particular. Chapter 4 will dive deep into where people got their information about the pandemic, comparing social media to other information sources such as mainstream media for the purposes of understanding facts about the virus and the ensuing circumstances. This chapter homes in on how people connected with others on such platforms through sharing and discussion of the pandemic.

In earlier work with Aaron Shaw, we investigated who edits the large online encyclopedia Wikipedia.[6] We proposed approaching the question of who participates in various online activities—of which sharing content about the pandemic on social media is one example—by thinking of online participation as a pipeline. The pipeline metaphor helps with recognizing that people have to pass through various points of the pipeline to remain "in the game" when it comes to active participation. In the case of Wikipedia, we detailed how, in order to arrive at the point of editing an entry on the encyclopedia, it is a necessary condition to be an Internet user, to have heard about Wikipedia, to have visited Wikipedia, and to know that anyone can make edits to it, before getting to the actual point of making a change to its content. In the case of social media participation, beyond being an Internet user, taking part in exchange about the pandemic on social media requires being signed up as a user of such platforms. Sharing content is often preconditioned on seeing content. To this end, before considering who was actively engaged with the pandemic on social network sites, the first step is to establish who uses them. It is also important to acknowledge that social media platforms vary and are not interchangeable, a point that the following section unpacks.

LARGE DIVERSITY AMONG SOCIAL MEDIA PLATFORMS

Although it is tempting to think about "social media" as a single milieu for communication, in reality these services make up a complex ecosystem of platforms that vary both in their affordances and their users.[7] A platform's *affordances* refer to what the system makes possible for users to do. It is easy to take features and options for granted as a user, but every reaction button, every posting option, every detail a user can see on a profile is the result of careful consideration on the part of those creating the service, a combination of engineers and, ideally, user-experience researchers.

For example, for many years, on Facebook it was only possible to follow someone's updates if two people had both agreed to be connected to each other, that is, to be "friends." In contrast, Instagram and Twitter had never had such reciprocity built into their systems. But they, too, differ in other important ways. While on Twitter it is easy to see who is following whom, including whether someone you are following follows you back, on

Instagram this is not at all self-evident, a design decision that the original founders of the photo-sharing service had made early in that platform's development.[8] Other such considerations include the ways a user can react to a post (e.g., through a like button, a heart, or a star); whether it is possible to edit something a user has posted (possible on Facebook, not on Twitter at the time of this writing) and whether to make this evident to the reader; or more recently, whether platforms decide to flag certain content when its accuracy is in dispute. There is, in fact, an entire domain of scholarly inquiry whose main focus is documenting in detail what apps allow and, in some cases, require (e.g., use of real names) users to do. For more on this, see the walkthrough method, an approach to using apps that can be informative to any user even if not implemented for scholarly purposes.[9]

Another important way in which social media differ is who tends to use them. Although there are some general trends in who more readily adopts social media (the most obvious being younger people), their users also tend to vary by gender, educational background, race and ethnicity, and metropolitan status, to name a few.[10] This is not usually due to constraints posed by the platforms themselves, rather, how they diffuse across the population. Some of this (as will become evident in the analyses below) is related to markers of digital inequality such as differentiated skills, but other variations have other reasons such as interest-driven user adoption.[11]

There is also considerable variation in which platforms are most popular across countries, which has important implications for cross-national comparisons. Perhaps the most apparent of these is how WhatsApp has taken the world by storm, except in the United States where it continues to trail behind many other services.[12] There are, of course, also changes over time as some sites completely fall off the map (only some readers will remember Friendster and MySpace)[13] while others gain prominence, in some cases, rather quickly. TikTok, for example, initially released in 2016, seemed to experience a meteoric rise in certain circles and regions during the pandemic, as did sites connecting neighbors like Nextdoor (although that had been around since 2008).

While generally popular, it is important to recognize that most social network sites are not universally used. Establishing who is left out of discussions that occur on such platforms is an important part of understanding who may benefit from them or be susceptible to their harms. Figure 3.1

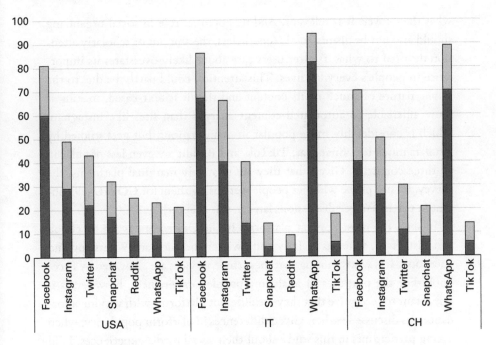

Figure 3.1
Percentage of respondents who use each social media platform daily (dark gray) and at all (light gray) in each of the three countries. (Reddit was not included on the Switzerland survey due to its low prevalence there.)

shows what portion of each country sample surveyed for this book was using the various platforms in April 2020, irrespective of COVID-19-related activities. Even a quick glance at the chart shows that there are considerable differences across the countries. In the United States, Facebook was the most popular, followed by Instagram and Twitter. In Switzerland and Italy, WhatsApp surpassed all others in popularity, followed by Facebook, and then Instagram followed by Twitter. Despite the amount of mass media coverage Twitter receives, it is worth noting that less than half of the population uses it in any of these three countries, figures that are similar to the site's relative popularity in other world regions as well.

That Twitter gets as much coverage as it does in the popular press is likely more a function of it being an important platform for journalists, and more recently some politicians, than its relative importance for the general population. Of course, insofar as journalists use it as a source for deciding what they cover, it is relevant. And its potential role in social organizing should also not be dismissed.[14] Nonetheless, the amount of scholarly attention devoted to what Twitter users care about likely overstates its importance in people's everyday lives. This attention could partly be due to the public nature of much of its content and that it is text-based, making it more amenable to automated content analysis than sites like Instagram, which is considerably more popular in many regions, but gets studied by scholars much less. Snapchat, TikTok, and Reddit are even less popular in all three countries. Given that they are relatively marginal platforms, the survey did not probe whether people were using them for COVID-19 content in particular (see discussions later in this chapter).

In sum, the most popular platforms in the United States are Facebook and Instagram, whereas in Italy and Switzerland they are WhatsApp and Facebook. It is worth noting that these three services are all owned by Facebook. For the purposes of analyzing people's experiences, however, it is important to recognize that they are different platforms with varying affordances, as discussed earlier. Given differences in platform popularity, when asking participants in this study about their social media experiences, I did not ask detailed usage questions about WhatsApp from American respondents, but focusing on it in the Swiss and Italian contexts was important.

When looking at the aggregate, the vast majority of the population uses one or another social network site in all three countries. In the United States, under 10% do not use any, among respondents in Italy, just 2% do not, in Switzerland, just 5% are not on any of the aforementioned platforms. In all three countries, older people are less likely to use social network sites. In the United States, there is also a gender difference, where women are more likely to use at least one such service. Higher general Internet skills are also linked to being a social media user.

WHO USES WHICH PLATFORM?

Before diving into the details of how people connected around topics related to the pandemic on social media, it is important to establish who used the various platforms in the first place during lockdown. By definition, those who are not on these platforms cannot be using them for exchanging information about the pandemic, so it is noteworthy to ascertain who those people are in the first place.

Several consistent patterns emerge across the three countries. In the majority of cases across the various social network sites, younger people are much more likely to be users than older people. In the United States, there is a clear association between socioeconomic status and social media use, where those with higher education and those with higher income are more likely to use services than others, a relationship that applies to every platform except Facebook. These associations are very different in Italy and Switzerland, however, where very few such relationships exist. (An exception is the case of Twitter in Italy, where those with higher education and higher income are more likely to have adopted that platform.) Disabled people are often more likely to be users of social media than people without disabilities, a relationship that is especially common in the United States context. Rural residence is often associated with lower levels of social network site adoption. Both higher Internet skills and more autonomy of use are often linked to higher rates of social media adoption. In the United States, Whites are less likely than most other racial and ethnic groups to be on social media even after taking into consideration different education and income levels.

Table 3.1 offers a summary of the results of logistic regression analyses that look at platform adoption by user background in the three countries. These analyses show how the likelihood of using the various services differs by age, gender, education, income, race/ethnicity (for the United States), rural residence, disability status, autonomy of Internet use, and Internet skills (all considered in the statistical models simultaneously). Consistent across all seven social network sites in the United States and most sites in Italy and Switzerland is that older people are less likely to be on them. (This age difference does not hold for Facebook in Switzerland and for Twitter in Italy.)

In the United States, women are more likely to be on Facebook, but less likely to be on Twitter, Reddit, WhatsApp, and TikTok. When education and income matter, they are associated with a higher likelihood of using sites (i.e., Instagram and Snapchat for income; Instagram, Twitter, Reddit and WhatsApp for education). Disabled people are more present than non-disabled people on four of the seven sites (i.e., on Twitter, Snapchat, Reddit, and WhatsApp). African Americans are more likely to use four services (Instagram, Twitter, WhatsApp, and TikTok), people of Hispanic origin are more likely to use Instagram and WhatsApp, and Asian Americans are more likely to adopt WhatsApp than Whites. Those who live in rural areas are less likely to be on WhatsApp.

Table 3.1
The sociodemographic factors, digital experiences, and skills associated with
using various social media platforms

Platform	Country	More likely to be a user—sociodemographics	More likely to be a user—user autonomy and skills
Facebook	US	Younger, female	Autonomy, Internet skills
	IT	Younger	—
	CH	—	—
Instagram	US	Younger, Black, Hispanic, higher income, more educated	Autonomy, Internet skills
	IT	Younger	Autonomy, Internet skills
	CH	Younger	—
Twitter	US	Younger, male, Black, more educated, disabled	Autonomy, Internet skills
	IT	Male, higher income, non-rural	Autonomy, Internet skills
	CH	Younger, male, more educated, disabled, non-rural	Internet skills
Snapchat	US	Younger, higher income, disabled	Autonomy, Internet skills
	IT	Younger, disabled	Internet skills
	CH	Younger, disabled, non-rural	—
WhatsApp	US	Younger, male, Black, Hispanic, Asian American, more educated, disabled, non-rural	Internet skills
	IT	Younger, nondisabled	Autonomy
	CH	Younger	Autonomy
TikTok	US	Younger, male, Black	Autonomy, Internet skills
	IT	Younger, disabled	—
	CH	Younger, disabled, non-rural	—
Reddit*	US	Younger, male, more educated, disabled	Internet skills
	IT	Younger, male, more educated, disabled, non-rural	Autonomy, Internet skills

* The Switzerland survey did not ask about Reddit owing to its low prevalence of use there.

Among study participants in Italy, there are fewer differences between users and nonusers of social network sites; nonetheless, adoption of these platforms is not random there either. Except for Twitter, age matters in all cases, with older people again less likely to use social media. Disabled people are less likely to use WhatsApp, but more likely to use Snapchat, TikTok, and Reddit. Women and those in rural regions are less likely to be on Twitter and Reddit. People with higher income are more likely to use Twitter.

There are even fewer patterns by sociodemographics in the case of Switzerland, at least concerning the three most popular platforms: WhatsApp, Facebook, and Instagram. The 70% of respondents who indicated using Facebook do not differ by any characteristic included in these analyses. This has important implications for using Facebook as a sampling frame for research in the Swiss context, because its users do not seem to bias for or against any particular sociodemographic background. In the case of WhatsApp and Instagram, older people are less likely to use them. Adoption of the other three sites (Twitter, Snapchat, TikTok) is more biased. Twitter is by far the most skewed in terms of its user base. Older people, women, and people who live in rural areas are considerably less likely to be on Twitter in Switzerland, while more educated people and disabled people are more likely to use it. These patterns concerning age, rural residence, and disability status extend to Snapchat and TikTok adoption as well.

When it comes to digital experiences, higher Internet skills are associated with more likelihood of platform adoption in the United States for all services examined here. Although it may be that using such platforms increases people's skills, since these are measures of general Internet skills and not social media skills, that is not necessarily what this finding reflects. My previous work has shown that Internet skills at a prior point in time explain social media adoption at a later point, so skills may well be part of the reason that people end up on various platforms in the first place rather than vice versa.[15] Autonomy of use also matters in that those with all three device access options (mobile, tablet, computer) are more likely to be on five of the seven platforms in the United States; only in the cases of the relatively unpopular WhatsApp and Reddit does autonomy not matter. These digital context factors also matter for the adoption of several platforms in Italy, although not for Facebook and TikTok. In Switzerland, digital experiences play a much smaller role in social media adoption, given that

Internet skills make a difference only in the case of Twitter, while more autonomy only matters for becoming a WhatsApp user.

Across the three countries, a few patterns emerge. Generally speaking, in almost all cases, older adults are less likely to be on social media. In almost all cases, having a disability correlates with higher propensity to use various platforms. When rural residence matters, it is related to lower likelihood of use. In cases where education or income matter, they are always associated with higher usage rates. The same is true for Internet skills and autonomy of use. Those with more privileged social and digital contexts are more likely to use social network sites. Insofar as there are benefits to be gained from such use, these are not equally distributed across the population, an important but oft-forgotten dimension of digital inequality.

ENGAGING WITH COVID-19-RELATED CONTENT ON SOCIAL MEDIA

The majority of respondents in all three countries reported using social media at least a few times a week (US: 68%, IT: 78%, CH: 60%) to communicate with friends and family who did not live in their household since the start of the pandemic, and many had done so daily (US: 44%, IT: 56%, CH: 35%). Several people noted that they had increased their use of social media since lockdowns (US: 34%, IT: 46%, CH: 26%). Given social media's broad appeal in providing easy access to others, it is understandable that people would turn to such communication technologies during lockdown. Especially at a time when everyone across the globe was experiencing something similar, it makes sense that people would try to find connection and information through such services. Of course, communicating with one's family and friends is but one of many activities one can do on social network sites.

One unique aspect of services like Facebook, WhatsApp, and Instagram is that they offer connections well beyond one's close ties, through professional and hobby interests, community affiliations, and other associations. They can also expose people to content through institutional accounts such as those of organizations to which people belong or various mainstream and independent news sources. Such information can reach people in numerous ways including organically through group membership, subscriptions, content one's connections share on the service, and advertising. Overall, social media are treasure troves of information with endlessly scrolling walls of material. Additionally, people may choose to go beyond the more passive behavior of looking at content and share material themselves to their networks. They can do this by resharing what others post or copying material from elsewhere, as well as by being the source of new content through such activities as asking or responding to the questions of others.

The figures presented in this section distinguish between *seeing* pandemic-related content as compared to *sharing* such material and to actively engaging with others through *discussion* to highlight the different ways in which people were interacting with others about COVID-19 on social network sites. In addition to describing general trends, this section highlights who was most likely to use social media for such purposes in the initial weeks of the pandemic.

Figures 3.2, 3.3, and 3.4 present the various types of pandemic-related content with which people engaged on social media across the three countries. The dark part of the bars indicates the portion of the population that *shared* such content, the full bars (the sum of the dark and lighter gray parts) represent those who *saw* such content. Understandably, more people saw pandemic-related material than shared it since sharing is often preconditioned on seeing such content in the first place (see discussion of the pipeline

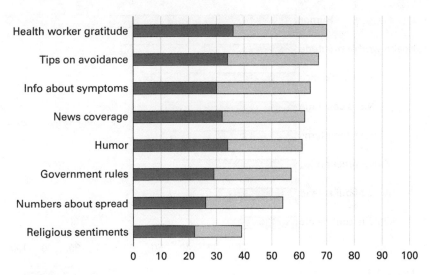

Figure 3.2
The relative popularity of various types of pandemic-related content people saw (full bar) and shared (dark part of the bar) on social media in the United States in descending order of popularity.

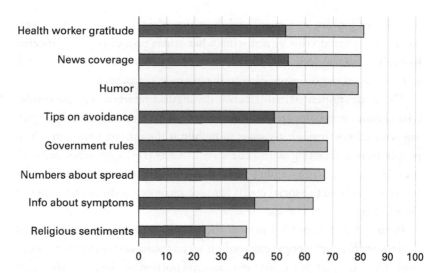

Figure 3.3
The relative popularity of various types of pandemic-related content people saw (full bar) and shared (dark part of the bar) on social media in Italy in descending order of popularity.

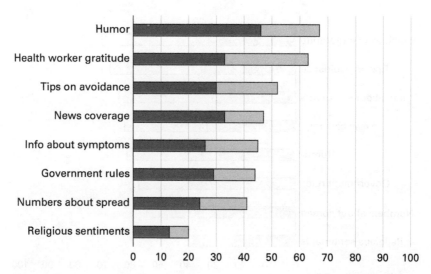

Figure 3.4
The relative popularity of various types of pandemic-related content people saw (full bar) and shared (dark part of the bar) on social media in Switzerland in descending order of popularity.

of online participation earlier in this chapter). Some aspects of these graphs generalize across the three countries, but other parts vary across the United States, Italy, and Switzerland.

In all three cases, expressing gratitude toward health workers is among the most common experiences both in the realm of seeing such content as well as sharing it. Seeing and sharing religious sentiments and teachings were the least common in all three countries. Also low on the list in all three are two more types of content: government rules about what people are allowed to do as well as numbers and charts about the virus's spread.

While both seeing and sharing information about the symptoms was relatively common among users in the United States, this was one of the least common types of content seen and shared by those in Italy, and it was in the middle of the pack in Switzerland. Seeing and sharing news coverage on social media was more popular in Italy than the other two nations. Humor was the most popular pandemic-related content in Switzerland, it was somewhat less common in Italy, and in the middle of the pack in the United States. Tips on how to avoid getting infected ranked in the top half in all three countries, highest in the United States then Switzerland and then Italy.

Overall, 78% of American respondents, 91% of Italian respondents, and 76% of Swiss respondents reported seeing at least one of these types of COVID-19-related content on social media. For sharing, these figures were 49%, 74%, 57%, respectively. Put another way, among those who saw any such content, 62%, over 80%, and 72% in the United States, Italy, and in Switzerland, respectively, also shared some such content. These numbers and all subsequent analyses in the chapter are based on the full samples. Of course only those who are social media users could have answered any of these questions in the affirmative, but I did not restrict the analyses to those subgroups because the overall question is who across the population engaged in such activities; thus taking the full samples as the baseline is meaningful.

Who was most likely to see such content in the three countries? In the United States, younger people were more exposed to material about the pandemic on social media as were women. African Americans, people of Hispanic origin, and disabled people also reported seeing such content more, but these results do not hold after controlling for other factors in the analyses such as education and income. More autonomy of use and higher Internet skills are also positively linked to seeing such content. In Italy, there are differences by age, but that is the only user characteristic that relates to having seen pandemic-related content on social media. The findings are similar in Switzerland, although those with higher income were less likely to report seeing such content than others when taking other factors into consideration.

The findings concerning who shared pandemic content on social media are similar in all three cases, with one notable difference. In both the United States and Switzerland, disabled people were more likely to share such content than others. Only in the United States does this finding hold when considering other factors in the statistical model, however. That is, in the Swiss case, once other aspects of socioeconomic status are held constant, there is no variation by disability status. In the United States, whereas 45% of people without disabilities reported sharing such content on social media, 70% of disabled people had done so—a considerable difference.

Beyond seeing and sharing content, on social media there is also the possibility of engaging others in conversation. Undoubtedly, many people had questions about the novel coronavirus, especially during the first few weeks of its global spread (the time of this data collection), so they were likely to turn to whatever sources were available to find and share answers and speculations. To capture this, the survey asked several questions about discussions related to COVID-19 on social network sites. These can be grouped into two types. One type concerned more general conversational behavior such as posting about one's own experiences, asking a question, answering a question, or receiving a response, participating in discussions about the pandemic, and correcting someone else's post or commentary about it. The other type concerned support—whether it was sought, offered, or received.

Figure 3.5 shows how common these two groups of engagement types were in the three countries. They happened at similar rates in the United States, where about a third of respondents had both actively discussed the pandemic and had exchanged more specifically around support-related content in the first few weeks of lockdowns on social media. In both Italy and Switzerland, participating in discussions was much more common than support exchange. The latter mirrors the rates in the United States, the former happened at higher levels in both countries, particularly in Italy, where over half of respondents had used social media to talk about the pandemic.

Interacting by way of more general discussion was considerably more common among disabled people compared to nondisabled people in all

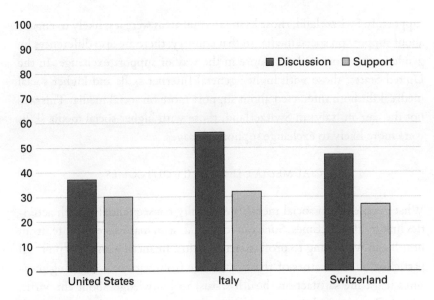

Figure 3.5
Percentage of respondents who participated in pandemic-related discussions and who asked for/offered/received support on social media in the three countries.

three countries (US: 52% among disabled vs. 35% among nondisabled, Italy: 68% vs. 55%, CH: 62% vs. 45%). In all three countries, young people were more likely to participate in such social media conversations. In the United States, women were less likely to engage in such discussions, but African Americans and those with higher education were more likely. In Italy (but not in Switzerland, in this case), those in rural areas were less likely to discuss the pandemic on social network sites. One more notable finding is that both general Internet skills and social media skills are positively linked with discussing the pandemic on social media in the United States and Italy (but not in Switzerland). Perhaps unsurprisingly, those who understand social media better took more advantage of engaging with others about the pandemic on such platforms.

Support exchange on social media was also more common among disabled people compared to their nondisabled counterparts in all three countries (US: 45% among disabled vs. 27% among nondisabled, Italy: 49% vs. 30%, CH: 44% vs. 25%). Also in all three, older people were less likely to have engaged in such support-related exchange. In the United States, those with higher education were more likely to ask for, offer, and receive support. In Switzerland, those living in rural areas were less likely to engage about support on social media. In that country, there are also differences by gender, with women doing more in the way of support exchange. In the United States, those with higher general Internet skills and higher social media skills both interacted about support more on social media. This was not the case in Italy; in Switzerland, those with higher social media skills were more likely to exchange support messages.

SOCIAL MEDIA USE AND LIFE OUTCOMES

What people do on social media is especially consequential if such activities link to life outcomes. Such outcomes can span many facets of life, from health and well-being to professional benefits, financial gains, and political participation. Indicators of these can range from various well-being measures (e.g., life satisfaction, health status) to knowledge about the virus. The challenge in making the connection between social media use and such outcomes is that they necessitate measures over time to see how changes in one are reflected in changes in the other. But since the pandemic was an unexpected event with extremely rapid changes in life circumstances, it is difficult to have well-being and other outcome measures from right before the pandemic to compare with measures collected once lockdowns started. This challenge concerns both before-and-after measures of online behavior as well as the types of life outcomes listed. Thus, the most we can do is determine whether certain outcome measures are correlated with social media activities without going too far in suggesting causal links. This is a classic example of the adage "correlation does not imply causation" and must be taken seriously. That is, the evidence presented here cannot speak to whether social media engagement about COVID-19 resulted in various changed life outcomes or whether people experiencing certain life conditions like lower life satisfaction were more likely to turn to social media to address their pandemic-related needs in the first place. Nonetheless, being able to document that such relationships existed between pandemic social media activities and life outcomes is helpful for understanding how people in different circumstances experienced lockdowns.

One finding from the data is that people who scored lower on a social connectedness measure—a measure that looks at people's feeling of belonging on a five-point scale—were more likely to engage actively on social

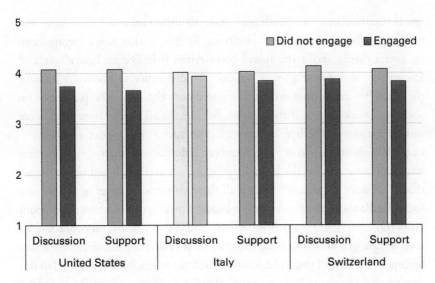

Figure 3.6
Level of social connectedness (*y*-axis, 1–5-point scale) between those who did not
engage on social media about the pandemic (light left bars) by way of general
interactions as well as support and those who did (darker right bars), across the three
countries. The lighter shades for discussion in Italy signal lack of statistical significance in difference.

media about the pandemic in all three countries. As figure 3.6 shows, this
difference is significant for support seeking/giving/receiving in all three
cases and for more general discussion in two of the three countries. It is
impossible to say, however, whether it is lower feelings of belonging that
led people to reach out on social media in this way or whether engaging
actively on social media then resulted in lower levels of social connectedness. For example, a person already feeling disconnected may have decided
to communicate with others on social media to feel less isolated. On the
other hand, it may also be that engaging with others in such a way resulted
in people feeling less connected if such interactions were negative in some
way either because they suggested to the user that they were worse off than
others or because their needs were not being met. Such experiences could
result in lower feelings of social connectedness. This work cannot address
what came first, but it does suggest topics for future research to investigate.

An important outcome woven through this book concerns people's level
of knowledge and misconceptions about COVID-19. Does engaging on

social media about the pandemic relate to differentiated know-how and misbeliefs about it? The overwhelming finding is that more engagement on social media about the novel coronavirus is linked to lower levels of knowledge about it. Figure 3.7 shows the difference in knowledge about the virus between those who participated actively about the pandemic on social network sites and those who did not. In all cases, those who were more engaged—whether by way of discussions or support exchange— knew less about the virus. The numbers indicate how many fewer correct answers people got on the knowledge score, on average, based on participation. The story is consistent across all three countries: engaging actively on social media about the pandemic is related to lower levels of knowledge about COVID-19. These results hold when controlling for several other factors such as age, gender, education, income, rural residence, and disability status (except in the case of general discussions in Italy where having engaged in this way on social media makes no statistical difference for pandemic knowledge).

In the interest of brevity, I aggregated different types of discussions about the pandemic for the purposes of analyzing the data, which are the

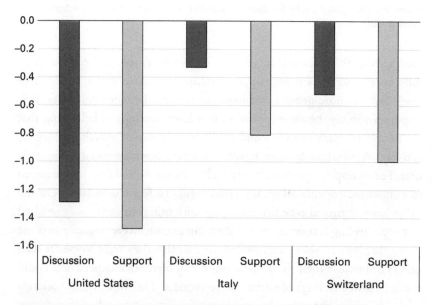

Figure 3.7
The difference in COVID-19 knowledge score between those who had actively engaged on social media about the pandemic and those who had not in the three countries. Active engagement is related to lower levels of pandemic knowledge.

analyses the previous paragraph discusses. It is worth highlighting the findings about one particular discussion type on its own, however. People who reported correcting others on social media about the pandemic had lower levels of knowledge about it (US: 10.0 knowledge score among those who did not correct others vs. 8.3 knowledge score among those who did, IT: 10.3 vs. 9.5, CH: 9.8 vs. 8.1). This is rather disconcerting since it suggests that those correcting others may be contributing to the problem of misinformation rather than addressing it.

Regarding misinformation about COVID-19, those who interacted with others on social media about the pandemic—whether by way of general discussion or in the form of support—were more likely to hold misconceptions about the virus such as that taking a hot bath (US: 16%, IT: 4%, CH: 4%) or drinking hot fluids (US: 21%, IT: 4%, CH: 9%) could help reduce risk of infection or that avoiding buying packages from China could do the same (US: 22%, IT: 3%, CH: 9%). (The appendix has all the details for all questions asked.) Figure 3.8 shows the extent to which active use of

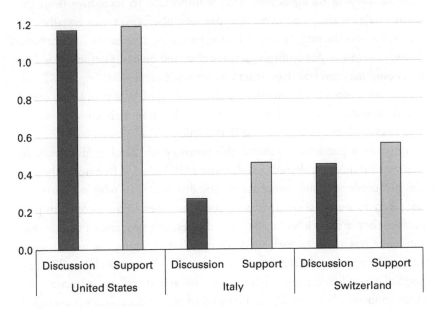

Figure 3.8
Difference in misinformation beliefs about COVID-19 between those who had actively engaged on social media about the pandemic and those who had not in the three countries. In all cases, active engagement is related to holding more misconceptions about the virus.

social network sites for coronavirus discussion and support, respectively, linked with more misinformation beliefs. This was especially pronounced in the United States, but also statistically significant in the other two countries. Although research from earlier years has shown that panics about the amount of fake news on social media are exaggerated, the complex political climate of 2020 in the United States may have contributed to the larger spread of misinformation there.[16] Similar to the case of differences in COVID-19 knowledge, controlling for sociodemographic factors does not alter the relationship of active engagement on social media and misbeliefs.

THE ROLE OF SOCIAL MEDIA ENGAGEMENT DURING LOCKDOWNS

The goal of this chapter was to focus on the public and semipublic discussions social media platforms make possible and probe how much people used such methods of communication to engage specifically about the pandemic. The chapter established that different social network sites attract users of varying backgrounds. This is important to recognize both for acknowledging that who decides to use such platforms is not equally distributed across the population and for appreciating that each platform has a distinct user base. Accordingly, actions observed on one platform (e.g., on Facebook) may not translate to actions observed on another (e.g., on Twitter), so for a more comprehensive understanding of how people use social media, it is important to gather data about behaviors on more than one social network site as was the case in this study.

Regarding pandemic content, the majority of social media users, in one way or another, shared and discussed COVID-19. But not only was this not universal, there were systematic differences in who was more or less likely to use these platforms for such purposes. Older people, in part because they are less likely to be on such sites, but not limited to that reason, were less active in such ways. In the United States, the more highly educated were more likely to interact about the virus on social media. The particular pandemic topics that people saw and shared varied across the three countries, but overall, the majority of respondents had some related experiences. Beyond sociodemographic differences, varied digital contexts also associated with divergent social media experiences whereby higher autonomy and higher digital skills both linked to more discussion and support exchange about the novel coronavirus on such platforms.

Disabled people were especially likely to share content about the pandemic. The findings about this group are worth extra consideration given this population's complicated relationship to digital media. As noted in chapter 2, disabled people seem to have caught up with others in their Internet skills. And as evidenced through the analyses in this chapter, during the pandemic, they were also more actively engaged on social media about it. But as the findings in the section focused on knowledge and misconceptions about the virus show, active online participation does not necessarily equate with more knowledge and fewer misbeliefs. In fact, the opposite relationship is true in this study: those who were more likely to engage about COVID-19 on social media understood the virus less and were more prone to believing misinformation about it.

Using social media to one's benefit requires that platforms be accessible for people of different backgrounds (including disabled people), but there is a history of accessibility problems on such platforms that seem to continue to pose challenges. An evaluation of five common social network sites in 2012 tested the accessibility of eight different components including keyboard shortcuts, color contrasts, and alt text for images.[17] LinkedIn scored the highest at 33%. Facebook scored 10%, and Twitter scored zero, meaning that it met none of the accessibility criteria. Such problems have led disabled people to develop their own workarounds such as Accessible YouTube, Easy Chirp, and You Describe to make the platforms easier to approach.[18] Some companies have put effort into developing more inclusive experiences, but many barriers persist.[19] For example, although Twitter offers an alt tag for images, many users either do not know about it or do not use it, given that many pictures posted on the site have no such tags to assist the visually impaired. Whether such accessibility issues explain why social media participation connects to undesirable outcomes for disabled people (e.g., their lower levels of knowledge about the virus and higher likelihood to believe misinformation) remains an important area for future work. Making sure that these platforms are inclusive of differently abled users should be a priority for these companies going forward, given their ubiquity and general importance for information exchange today.

In sum, it is clear from the evidence discussed in this chapter that many people turned to social media to engage about the pandemic. Whether this was ultimately for the better is unclear based on the results presented in this chapter. Although such engagement was linked with lower levels of social connectedness, less knowledge about COVID-19, and more misinformation beliefs about it, the evidence does not allow for making causal claims. It does raise important questions, however, for future work about the repercussions of engaging with others on social media in unsettled times.

While this chapter mainly focused on interactions, the next chapter examines the role such platforms played in how people informed themselves about the virus and the societal changes it precipitated in the context of other information sources such as traditional media and varied online resources. It puts social media use in the context of other information sources—from television to health websites and from newspapers to online government information—to see how these comparatively informed people in their efforts to learn about the pandemic.

INFORMATION SOURCES AND
(MIS)UNDERSTANDING COVID-19

Rarely is access to reliable information as important as during a global health crisis. The World Health Organization (WHO) became so concerned about the overwhelming amounts of information circulating globally about the novel coronavirus—some of it correct, some misleading—that it organized the first WHO Infodemiology Conference in June 2020 to start addressing these issues.[1] Chapter 3 examined how people were discussing the pandemic on social media, but it did not consider how they were getting information about COVID-19 in the broader media ecosystem. Participating in social media discussions was, of course, just one way of gaining an understanding of the situation in the vast media environment. To fill out the bigger picture, therefore, this chapter looks at the broader media context of how people learned about the pandemic. It examines what information sources people were using from traditional channels to online-only resources and how this related to their knowledge and misconceptions about COVID-19.

There is a long history in communication scholarship of studying whether access to and use of various media sources benefit people's knowledge levels differently based on their socioeconomic status—the so-called knowledge gap hypothesis. There is also considerable work on both credibility assessment and the diffusion of misinformation. Curiously, these two areas of research rarely intersect. The next sections give some background of this work in order to show why information sources are important to consider when examining how informed and possibly misled people were about the pandemic, and how this relates to the problem of inequality. At a time when having the right knowledge can mean the difference between life and death, such information inequalities are of paramount significance.

THE KNOWLEDGE GAP HYPOTHESIS

In the 1970s, communication scholars proposed the "knowledge gap hypothesis" concerning the diffusion of knowledge across the population as more information is made available through different media.[2] Its core idea is that as more content is available, those of higher socioeconomic status will benefit from the increased number and variety of resources disproportionately compared to those from less privileged backgrounds, implying that the gap between the information poor and information rich will widen over time rather than decrease. There are several reasons why this may be the case. At the most basic level, the theory goes, those who access and engage with more information will benefit from this across different domains of their lives, from doing better in school to being more informed, for instance, about politics or about health matters. Over time, these people will learn to acquire information more quickly, thereby facilitating even more knowledge gains in the long run and thus contributing to a spiral of advantages.

Over the decades, hundreds if not thousands of studies have tested for knowledge gaps across different countries concerning varying topics (e.g., news, science) and diverse media settings (e.g., television, newspapers, the Internet).[3] Indeed, this research focus has been so popular that several papers exist whose sole purpose is to aggregate and summarize what hundreds of these studies have found.[4] It is beyond the scope of this book to review this literature in detail, but the following pages highlight some of its key findings. Generally speaking, work has often noted that education level explains differences in knowledge obtained through varying media sources. And yet, whether and to what extent those with more education reap more rewards from the sources they use than those with less education is not as straightforward.

The now world-famous children's television show, *Sesame Street*, debuted in 1965 as part of US President Lyndon B. Johnson's Great Society initiative.[5] Its goal was to close the gaps between children of different means by providing free educational programming to the more than 92% of American households that had a television set at the time, a number that has since increased.[6] But tests of the show's impact revealed already after its first year that children who watched it more gained more from it, and that these tended to be children from relatively privileged households.[7] This provided support for the knowledge gap hypothesis in that those from higher socioeconomic status were gaining more from the available educational material than their less privileged counterparts. To be sure, all children who watched the show gained knowledge, but the more privileged did so at higher rates.

Among adults, television is less likely to be associated with knowledge gaps, however, perhaps because the main networks (ABC, CBS, and NBC in the United States; SRG in Switzerland; RAI in Italy) broadcast relatively standardized content targeting large socially diverse audiences.[8] Indeed, a study looking at political knowledge during the 1996 US presidential campaign found that the knowledge gap between high- and low-educated people was smaller for those of heavy TV consumption than those of light consumption.[9] A study examining political knowledge gaps based on several Norwegian election studies found that TV tends to stabilize while newspapers tend to amplify such gaps.[10] This finding is also the general conclusion of the most recent and comprehensive meta-analysis of this scholarship to date showing that television maintains knowledge gaps, while print media increases them by educational background.[11] Much less research has focused on radio, and the little that has, has not found any significant gap-widening effects.

KNOWLEDGE GAPS IN THE INTERNET AGE

As the Internet gained widespread popularity, research also started looking at how online media may influence knowledge gaps.[12] Theoretically speaking, the Internet can have two opposing influences on people's information acquisition, or have none at all. On the one hand, it can offer easier access at lower cost to a lot of content, allowing users to devote less time and effort and also fewer resources to gathering information. On the other hand, if the higher-educated use digital media for informational purposes more than the lower educated, this could widen knowledge gaps. Indeed, considerable research across countries and over time has found this to be the case: people from higher socioeconomic status are more likely to use digital media for informational purposes than those from less privileged backgrounds.[13] This would suggest the potential for increased knowledge gaps or a reinforcement of the status quo of inequality by social position. Similarly, if those who are more skilled with digital media take better advantage of online resources, this too could contribute to knowledge gaps.[14] A third possibility is that, given its myriad of information types and sources, the Internet does not in and of itself make a difference for how people of varying socioeconomic status obtain information through it; that is, off-line gaps and the resulting tendencies are simply mirrored online.

Some work has tested the knowledge gap hypothesis comparing Internet use with traditional media. A study of Dutch adults collected data at three points in time two months apart to examine how television, newspaper, and news website usage related to knowledge of current affairs.[15] In line with results mentioned in the previous section, television viewing was linked to higher knowledge and those from lower educational backgrounds gained more through this type of media use. Newspaper use also increased knowledge, but there was no difference here according to education level. The effect of online news sources mirrored those of newspaper reading: gain in knowledge over time with more use, but no difference by education, suggesting no gap-widening effect.

A study of Flemish media users in Belgium over several weeks found that traditional media consumption was positively linked with current affairs knowledge, while the use of Facebook was not.[16] The study did not consider differences by educational background, however, so it could not speak to whether or not knowledge acquisition varied by educational level across information sources. A study looking at Americans' exposure to online news from 1998 to 2012 found that, over time, more such exposure happened across different socioeconomic groups and was linked to reducing political knowledge gaps.[17] In sum, results seem mixed, which is perhaps not surprising given the different time periods studies investigate and the changing nature of online resources and the media ecosystem more generally over the years.[18]

Today, many people are exposed to news online via social media channels, whether this exposure is deliberate or incidental, that is, coming across it without a concerted effort to see it.[19] In the earlier days of the Internet, online news consumption was considered a "pull" activity, in which users saw online news when they actively sought it out themselves.[20] News consumed through social media is often more of a "push" experience, given that users receive non-searched-for information interspersed with more personal and social content from their networks.[21]

A study analyzing national panel survey data in the United States collected in 2008 and 2010 found that more Facebook use associated with more political knowledge for highly educated people, but not for those with less education.[22] A big challenge of such studies is that their findings may not hold up over time as platform affordances change, as do their user bases, and correspondingly the types of content that gets shared on them (see chapter 3 for more on these points). Given the large amounts of COVID-19 information (including misinformation) circulated on social media, the question of whether a reliance on such platforms to learn about the pandemic links to socioeconomic background is worthwhile.

As evidenced by the body of work reviewed above, much of the knowledge gap literature focuses on news and current events mainly in the political realm. Nonetheless, several studies have also addressed health and science knowledge.[23] Analyses from the Internet's early mass diffusion years suggest its potential to help with reducing knowledge gaps, but as noted above, how this holds up as the Internet has matured and social media have proliferated remains an open question.[24] Also notable is that past work has found variations across countries in people's understanding of science matters, so a comparative angle can be helpful for appreciating whether national differences exist in people's understanding of the novel coronavirus.[25]

There is even less prior work specifically focused on knowledge gaps during crisis events such as epidemics and natural disasters. Researchers looked at people's information seeking about and preparedness for a hurricane in Texas and found no differences by education or income in either.[26] Evaluating Singaporeans' H1N1 influenza knowledge, one study found that newspaper reading was not associated with knowledge gaps, while television viewing reduced them.[27] That article also noted the population's high level of trust in the media for other health crises, which it argued may have helped people take the government's public communication strategy seriously.[28] These differing findings suggest that there is a need for more work on knowledge gaps and the role of varying information sources in unsettled times.

A particular concern during the pandemic was the widespread diffusion of misinformation, or the oft-popular term for it, *fake news*.[29] Whether spread with malicious intent, out of fear, or simply a misunderstanding, misconceptions about health matters can have real-life consequences and are thus worthy of their own investigation. The Pew Research Center found that the majority of Americans who get most of their news during initial lockdowns from social media reported seeing fake news about the novel coronavirus already during the early days of outbreaks in the United States.[30] While measuring who can identify misinformation is not the equivalent of knowing how much false information is out there, it is one way to get at whether people were encountering potentially questionable content about the virus from early on in the pandemic, which the report suggests they were. The presence of misinformation on social media is certainly nothing new, although it is not necessarily shared or believed as readily as public rhetoric might suggest.[31]

Few studies have explicitly investigated the knowledge gap hypothesis as pertaining to misinformation. Reporting on data about people's online search habits from seven countries (including two of the countries that are the focus of this book, Italy and the United States), one article noted that higher income and education were related to being less vulnerable to experiencing information problems, although it also emphasized that people from across the socioeconomic spectrum can be exposed to problematic content.[32] The study did not consider, however, whether these vulnerabilities would vary by type of information source. Generally speaking, most work either focuses on knowledge or on misinformation beliefs, but rarely does literature bring the two together. This book is unique in that it considers both.

The COVID-19 pandemic offers a valuable opportunity to test both how varying information sources link to knowledge differences as well as vulnerability to misinformation in crisis times concerning a topic with very real consequences for most people. There was an acute need for people to understand how to stay safe so identifying what sources correlated with such knowledge is significant. Concurrently, since plenty of misinformation circulated about how to avoid the deadly virus, examining what sources related to misconceptions is also relevant to this particular calamity. Given the considerable efforts that many governments put toward informing the public and because we know from chapter 1 that some people nonetheless lacked knowledge, these examinations can offer lessons for how best to reach various populations with crucial messaging.

DIGITAL SKILLS AND KNOWLEDGE GAPS

In the digital media environment, an important factor that could contribute to knowledge gaps concerns people's digital skills. Chapter 2 showed that people in this study varied in their general Internet skills as well as their social media skills, often reinforcing existing social inequalities. Understanding digital media better may help people identify credible sources and sidestep questionable ones. It can also help them home in on desired and applicable content through more refined searches. Work on information seeking has shown that those with higher Internet skills can find content better, which may be particularly relevant at the time of a health crisis with confusing information circulating about it.[33]

While researching who makes editorial contributions to the large open online encyclopedia, Wikipedia, my collaborator Aaron Shaw and I found that a major barrier to such contributions was knowing that anyone could edit the site in the first place.[34] Who was more likely to know that Wikipedia can be edited by anyone? People with higher Internet skills. And more educated Internet users with higher skills were considerably more likely to edit the site than less educated Internet users with higher skills, finding evidence of a knowledge gap.

Another way that digital skills may matter is in knowing how to deal with information overload, something that is especially relevant during the COVID-19 lockdown period.[35] Indeed, an April 2020 report from the Pew Research Center showed that the majority of Americans found the situation emotionally taxing and needed to take breaks from the news already in the early months of the pandemic.[36] From this perspective, however, more information sources and media use may not translate to better knowledge because the latter may partly depend on the ability to extract relevant and reliable information at a time of high stress and a deluge of information.

The main takeaway from the knowledge gap literature for the question of how people used digital media during lockdown is to recognize that information sources and what people gain from them may vary by sociodemographic background. To see how this played out during the pandemic's first months, the sections that follow look at what information sources people used to learn about COVID-19, how these varied by user background, and how people's knowledge and misperceptions about the virus related to where they got their information. It does so by also considering the role of autonomy of Internet use (number of Internet access devices) and digital skills in the process. But first, it is helpful to establish who was following news about COVID-19.

FOLLOWING NEWS ABOUT THE PANDEMIC

Before looking at where people got information about the novel coronavirus during initial lockdowns, it is useful to ascertain how closely people were following related news in the first place. Figure 4.1 shows the answer from each of the three countries to the question: "How closely, if at all, have you been following news about the outbreak of the Coronavirus also known as COVID-19?" Clearly, the vast majority of people were following it to some extent, with over half in each following it very closely. The Swiss were the least likely to pay close attention, although the proportion of those who did not follow the news at all or not too closely is similar to the United States at 6%. Italians were the most engaged with pandemic news with only 2%

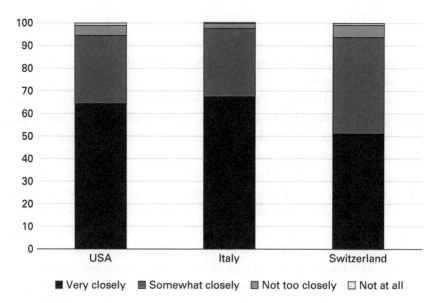

Figure 4.1
How closely people followed news about the novel coronavirus outbreak in the three
countries.

claiming they did not follow it at least somewhat. This may be due to the
severity of lockdowns there compared to the other two countries and also
the more extreme spread of the virus in Italy during those early weeks.

Those more likely to follow news closely about COVID-19 varies across
the countries somewhat, although there are similarities. People with higher
incomes in all three were more likely to follow such news (although in Italy,
only marginally). In the United States and Switzerland, this was also true of
those with a university degree. Why this was the case could have several expla-
nations. Those with higher incomes and higher education were more likely to
be in occupations that would move to working from home, thus disrupting
daily work routines and a need to keep abreast of shifting circumstances. On
the other hand, those in jobs that could not be shifted to home such as restau-
rant staff or those in the hospitality industry had a greater chance of being laid
off or at minimum losing some of their income, so news about the pandemic
was certainly relevant to their circumstances. Of course, those in such precarious positions may have been so busy trying to figure out how to find alternative sources of income that they did not have time to follow the pandemic's
course closely.

Regarding age, in the United States, those under thirty were less likely to follow COVID-19 news, but there were no age differences beyond that. In Switzerland, this change in interest kicked in for those under forty, with no reported difference among those in their forties and older, all of whom followed it more than younger adults. Since it became clear early on that older people were more severely affected by the virus, this may explain lower interest among younger cohorts. In Italy, there is a U-shaped relationship between age and following pandemic news, where those in their forties, fifties, and sixties followed the news more than young adults and those seventy and older. There were also racial differences in the United States, with Blacks following more than Whites when holding other sociodemographic factors constant. This may be due to the complicated history of health care for many African Americans in that country, as detailed in chapter 1.

In a separate analysis, I also considered how political leaning linked to following pandemic news. In the United States and Switzerland, those on the right were less likely to follow closely, but in Italy there were no differences by political orientation. Once this variable was added to the full statistical model with the sociodemographic variables, in the United States, Blacks were no longer more likely to follow COVID-19 news than Whites, suggesting that the racial finding there was being driven by African Americans' larger propensity to lean Democrat. In Switzerland, the addition of political orientation to the model results in political leaning itself no longer relating to propensity to follow news about the novel coronavirus. The overall image that emerges is that, perhaps unsurprisingly, a few weeks into lockdowns, most people were curious to keep up with pandemic goings on.

Did following COVID-19 news very closely link to being more knowledgeable about the pandemic? Yes. As figure 4.2 shows, those who did so had better knowledge about it, all statistically significant differences. This interest in COVID-19 news, however, did not translate to holding different levels of misconceptions about the virus (in the case of Italy, there is a marginal difference, where those who followed pandemic news very closely were somewhat less likely to hold misinformation beliefs). How specific information sources that people consulted and relied on for news about COVID-19 may have played a role in people's knowledge and misbeliefs about the virus is the focus of the rest of this chapter.

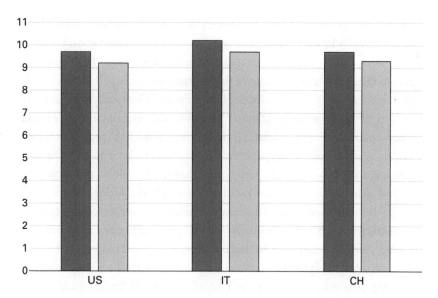

Figure 4.2
People's knowledge about COVID-19 by following pandemic news very closely
(darker left columns) versus less so (lighter right columns) by country.

INFORMATION SOURCES ABOUT COVID-19

As countries shut down from Italy's severe lockdown to Switzerland's softer
one, many people found themselves in a situation they had never experienced before: physically isolated from all but their cohabitants. Such unprecedented circumstances with uncertain next steps and little known about the
lethal virus understandably led many people to seek out information about
the pandemic. This section looks at the various media sources they consulted
ranging from more traditional media to online-only resources.

The survey asked about public and commercial television (in the United
States, both the major networks and cable), radio, magazines, newspapers,
late-night talk shows, social media, and various online-only resources. (The
appendix lists the precise wording of the questions and media included
for all three countries.) In the case of traditional media (television, radio,
paper), respondents were asked to report on consulting both off-line and
online sources. The online-only sources included YouTube (excluding videos from mainstream news outlets), online-only news media, government
websites, health websites, websites displaying numbers about the global

spread of the virus, and those displaying numbers about its spread in the country. The survey did not give examples of these categories nor distinguish among sites that corresponded to them. That is, if someone indicated having relied on health websites as a source of information about the pandemic, it is not possible to distinguish between a health website written and hosted by medical professionals at a health research facility versus a health blog maintained by someone who eschews vaccination of all types. It was up to the respondent to interpret the category as they saw fit. This means that it is impossible to know whether, for example, the health websites they may have consulted were reputable and trustworthy or not.

It is also not possible to know what type of content people were seeing on the various sites. This is particularly limiting in the case of social media because the sources of information on such platforms can vary considerably. The survey does not distinguish between seeing friends' posts on Facebook or following the Twitter account of a government agency. As the functions of social media increasingly encompass most online options—from exposure to family members' commentary to formal news sources—it becomes less useful to consider them as one uniform information source. This is a limitation of the data collection approach here, but it was necessary in order to be able to ask the many other questions analyzed throughout the book. Despite these shortcomings, it is nonetheless insightful to be able to identify in what online contexts people were learning about the rapidly evolving global pandemic because that can shed light on what types of interventions may be most useful in the future to ensure more knowledge and fewer misconceptions.

Figure 4.3 shows the relative popularity of media types across the three countries. Overall, across all three countries, television and online-only sources (excluding social media [SM], which is its own separate category) were the two most popular media categories. In the United States and Italy, social media as a category followed online-only sources, and in Switzerland traditional newspapers and magazines (whether online or off-line) came in third. In all three countries, radio was the least popular medium, although it was still consulted by about half of respondents, if not more. Already at this basic level of analysis, country variations arise, which is an important reminder to be careful about generalizing findings from one national context to another and to conduct comparative analyses whenever possible.

Figures 4.4, 4.5, and 4.6 show in decreasing order of popularity the various information sources by media type in each country. In the United

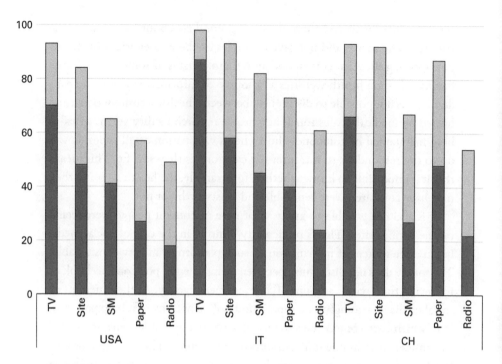

Figure 4.3
The relative popularity of media types as information sources about the pandemic across the three countries.

States (figure 4.4), the three major traditional commercial broadcast network channels (ABC, CBS, NBC) were by far the most viewed television sources compared to cable news channels. The survey did not ask people whether they were viewing local or national news programs on the network channels, but it is helpful to keep in mind that they often have both and that these follow each other on the schedule, which may encourage more viewership once someone is tuned into a broadcast. (In the age of digital on-demand television, scheduling may be less relevant than it once was; for keeping abreast of ever-changing circumstances amid a global pandemic while constrained to one's home for the most part, there is a good chance that many people were watching news shows live, and so local and national news following each other may have resulted in people watching both.)

Several cable news channels were also popular. Some of these are especially slanted toward certain political ideologies (Fox leans right, MSNBC leans left, with American CNN in between). Late-night talk shows also

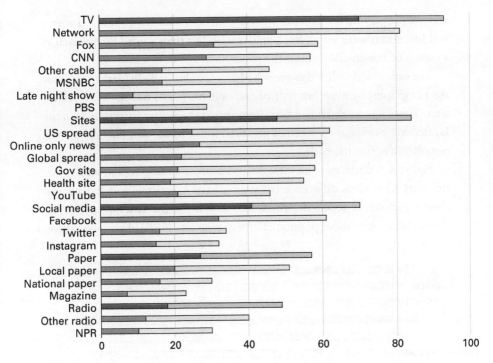

Figure 4.4
Information sources about the pandemic among respondents in the United States
grouped by media type: TV, websites, social media, paper, and radio; darker part of
the line refers to a source consulted daily or almost daily, and the lighter part to a few
times a week or less.

often discuss politics, which research has shown can influence people's
political knowledge.[37] However, less is known about whether late-night
talk shows influence other areas of understanding, such as those concerning
health topics. In the United States, public media is relatively weak (PBS
television and NPR radio networks, respectively), which explains their
relative lack of popularity shown on figure 4.4.

People consulted different online-only resources for keeping abreast of
the pandemic. More than half of Americans reported consulting websites
displaying numbers about the spread of the virus in the United States and
globally (an example of this would be Worldometer), looking at government and health websites, and online-only news sources for pandemic
information. On a daily or almost daily basis, Facebook was a more common source than most of the above, surpassed only by network television

and sites displaying numbers about the virus' spread in the country. Twitter and Instagram were much less popular with less than a third listing each as a source of information about the pandemic.

The only resource listed as consulted daily or almost daily (dark portion of the bar graphs) by even just half of respondents (48%) was network television. These figures show just how fragmented the media ecosystem is in the United States even for a topic of great relevance to the vast majority of the population across the nation.

Figure 4.5 illustrates the information sources among Italian study participants. One clear difference between where Americans and Italians got their information about the pandemic is that for the latter respondents, most sources were more popular. The vast majority (98%) of the population

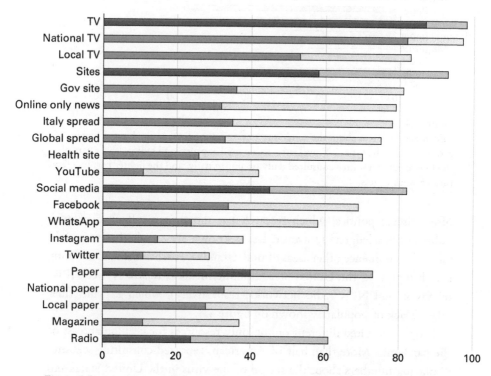

Figure 4.5
Information sources about the pandemic among respondents in Italy grouped by media type: TV, websites, social media, paper, and radio; darker part of the line refers to a source consulted daily or almost daily, and the lighter part to a few times a week or less.

watched national television with more than 80% doing so daily or almost daily with local television not far behind.

Online-only resources were also very important, with government sites leading the pack. Indeed, in Italy, more than three-quarters of people consulted most site types for information, and a third did so daily or almost daily. There, unlike in the United States, Facebook did not surpass websites in popularity. WhatsApp was a source for over half of Italians, with Instagram and especially Twitter far behind. Just under three-quarters of Italians relied on newspapers, whether national or local, to stay informed about COVID-19.

In Switzerland, the Swiss Broadcasting Corporation (SRG SSR on figure 4.6) is a public broadcasting association paid for by annual fees

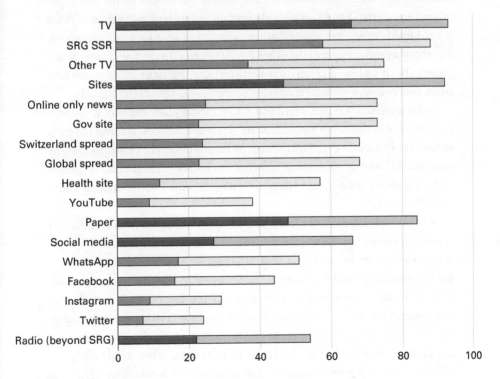

Figure 4.6
Information sources about the pandemic among respondents in Switzerland grouped by media type: TV, websites, paper, social media, and radio; darker part of the line refers to a source consulted daily or almost daily, and the lighter part to a few times a week or less.

collected directly from all households. It offers television, radio, and online content in German, French, and Italian across the country's various language regions in addition to seven more languages online. The survey asked about all of these modes of information dissemination together, so it reflects not only television viewing, but all properties of SRG SSR. (It is listed under TV on the graph because it runs several channels with news shows that are very popular.) Not surprisingly, SRG is the single most popular source of COVID-19 information, with 89% of the Swiss having consulted it and 59% having done so daily or almost daily.

Swiss respondents relied on online-only resources about as much as they did television, with online-only news sources and government sites consulted by 73% each. Following numbers about the spread of the virus specific to Switzerland and also globally was also popular at more than 67% for both. Health sites were somewhat less accessed as an informational source, but about as common as in the other two countries. In the Swiss case, newspapers (both online and off-line versions) are the next most important pandemic resource with 87% consulting them and 48% doing so daily.

In Switzerland, no one particular social media platform was as broadly consulted as in the other two countries. Unlike Facebook leading the pack in the other two cases, in Switzerland, WhatsApp was the platform that the most (51%) listed as a pandemic information source, 17% as daily or almost daily. Facebook follows at 44%, with Instagram and Twitter trailing at less than 25% each.

Were people of different backgrounds more likely to turn to different types of information sources for keeping abreast of the pandemic? Table 4.1 shows the results of statistical analyses (logistic regression) that accounts for the various sociodemographic factors listed in the first column simultaneously when considering what explains relying on various information sources about the pandemic. What this means is that people of higher income, for example, were more likely to have turned to most sources (excepting social media) in the United States even when controlling for all other user background characteristics. Consistent across the three countries is that younger adults were more likely to list online-only resources and social media as pandemic information sources. In general, these results show that people of different backgrounds relied on varying sources to learn about the virus. It is also worth noting, however, that in some cases there are no differences. For example, in the United States, there is no

Table 4.1
The relationship of sociodemographics and information sources for learning about the pandemic in the three countries

User	Country	TV	Site	Social media	Paper	Radio
Younger	US		+	+		+
	IT		+	+		
	CH		+	+		
Women	US			+	−	−
	IT				−	−
	CH					
Higher education	US		+		+	+
	IT	−			+	
	CH	−				−
Higher income	US	+	+		+	+
	IT					
	CH	+				
Rural	US					−
	IT	−				−
	CH	−				
Disabled	US			+	+	+
	IT					
	CH			+		+
Hispanic	US					
Black	US	+		+	+	
Asian	US					
Native	US	−				

Note: Plus sign denotes positive association, minus sign denotes negative association, blank cell denotes no association between the user characteristic and the particular media type.

difference between people of Hispanic and Asian background as compared to Whites; in Italy, there is no variation by income or disability status; and in Switzerland, there is no difference by gender.

Media sources can have political leanings, so it is also worthwhile to examine whether there are differences by political orientation in who consulted what sources. In the United States, the only media where this comes up concerns television and traditional newspapers. In particular, when controlling for sociodemographics, those leaning more Republican are less likely to have relied on the three main commercial networks (ABC, CBS, NBC), CNN, MSNBC, and PBS for news about the novel coronavirus, but they were much more likely to have watched Fox News. They were also less likely to list national newspapers as a source (but there is no ideological difference for turning to local papers). In Italy, the only difference for television concerns talk shows, which those leaning to the right were more likely to list as a source. This group was also more likely to turn to Facebook and YouTube for COVID-19 information. In Switzerland, those leaning right were more likely to list Facebook and WhatsApp as information sources, as well as free (e.g., 20 Minuten/20 Minutes) and tabloid papers (e.g., Blick).

On the whole, what information sources were most popular across the three countries differed as did how many people consulted particular sources to learn about COVID-19. Beyond television (especially the three main traditional commercial broadcasters in the United States, although even there a political divide seems to exist, and the national networks in Italy and Switzerland), there is quite a bit of variation in where people got their pandemic information. As noted already for the US case, this fragmentation of attention shows just how varied the media ecosystem has become in the twenty-first century. Even when people are faced with imminent and generalized danger, where they get their information varies widely. How do these different sources then translate into knowledge as well as misinformation beliefs about the novel coronavirus?

THE LINK BETWEEN INFORMATION SOURCES
AND COVID-19 KNOWLEDGE

To determine whether consulting a particular information source is related to understanding the pandemic, I control for both sociodemographics as well as relying on other information sources all in the same statistical model. This approach is preferable to comparing just the information source with the knowledge score as it considers the fact that different types of people access different media types. That is, were we to find that a particular source relates to higher knowledge, it may simply reflect that more educated people are more likely to rely on that source and it is ultimately education that explains the difference in knowledge, not the information source. Accounting for multiple channels simultaneously is important as learning about the virus did not happen in a vacuum removed from other potential sources.

Relying daily on commercial network news for information about the novel coronavirus is associated with more pandemic knowledge in the United States (doing so less than daily compared to never is not). In Italy, watching news on national TV stations daily is associated with higher pandemic knowledge. In Switzerland, SRG as a daily information source is positively linked to COVID-19 knowledge. In all three countries, these relationships hold regardless of education level, suggesting that there is neither a widening nor a closing knowledge gap.

Few other information sources link to more knowledge. In the United States, following the spread of the virus in the country relates to more pandemic know-how. This is also true in Switzerland, where relying on online-only news sources and government sites is also associated with more virus knowledge. In Italy, no other source exhibits this positive link.

In all three countries there are some information sources that negatively link to understanding the pandemic. In the United States, watching MSNBC or PBS daily relates to less knowledge, as do magazines and Instagram as information sources. In Italy, multiple sources show this negative association, including occasional radio listening, daily newspaper reading, and relying on Instagram or YouTube. Several sources relate to lower knowledge levels in Switzerland as well: occasional radio listening, relying on Facebook or Twitter occasionally as an information source, relying daily on WhatsApp or YouTube, or at any level on Instagram. Curiously, listing health sites as an information source also links to lower pandemic knowledge in this country.

Consulting various social media platforms about the pandemic often associated with less knowledge. Might social media skills make a difference in this relationship? Figure 4.7 shows the relationship of social media skills to COVID-19 knowledge by whether a person consulted Twitter as a source of information about the pandemic. (These are based on results of regression analyses that account for various sociodemographic characteristics.) Lines with the same shade (black, gray, light gray) depict the same country (the United States, Italy, and Switzerland, respectively). The solid lines refer to people who reported not using Twitter as an information source to learn about the pandemic, the dashed lines refer to people who did. In the case of the United States and Switzerland, people's pandemic knowledge does not vary by social media skills among those who did not

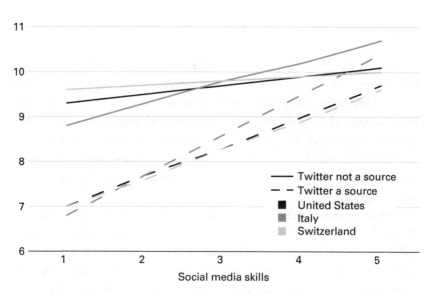

Figure 4.7
The relationship between COVID-19 knowledge (*y*-axis) and social media skills by
whether the respondent used Twitter (solid line = no; dashed line = yes) as an
information source about the pandemic; black lines = US; gray lines = Italy; light gray
lines = Switzerland.

rely on Twitter for COVID-19 information, which makes sense since they
are less dependent on social media for content. However, among those who
did consult Twitter, being more skilled links to higher knowledge about the
virus. These findings suggest a gap between the more and less skilled in how
much benefit they may reap from using the platform. Note, however, that
compared to those who were not relying on Twitter for novel coronavirus
information, those who were had lower pandemic knowledge.

In the United States in particular, most information sources link to less
knowledge. How is this possible? It is important to remember that people's
exposure to information occurs across different sources, that is, exposure
from one source (e.g., watching television) does not happen in isolation of
exposure from another (e.g., using Facebook). In the United States, after a
certain number of media types, the more sources people consulted for pandemic information, the less they knew about the virus. Figure 4.8 shows
this relationship for all three countries. In the United States (and to a much
lesser extent in Italy), there is a reverse U-shaped relationship between
COVID-19 knowledge (*y*-axis) and number of information source types.

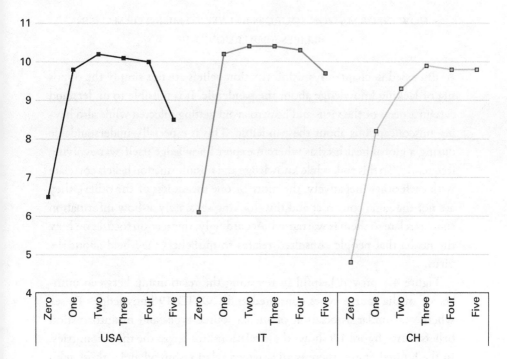

Figure 4.8
Level of COVID-19 knowledge (*y*-axis) by number of media types consulted (*x*-axis) in the three countries.

This signals that relying on very few source types is associated with less knowledge, but so is relying on many source types. It looks like, at least in the US context, turning to too many sources may have caused too much confusion at a time when much uncertainty still remained about the virus.

It is notable that including information sources in the statistical models did not change the relevance of sociodemographic characteristics that linked to more or less COVID-19 knowledge, as discussed in chapter 1. Although the addition of information sources strengthens the statistical models in that they explain more of the variance in knowledge, they do not explain the variations by age, gender, race, education, income, and disability status that the earlier analyses had uncovered. The results in this section are also robust to the inclusion of digital inequality indicators such as autonomy of use and social media skills. Additionally, they all took political leaning into account.

HOW INFORMATION SOURCES RELATE TO MISINFORMATION
BELIEFS ABOUT COVID-19

As discussed in chapter 1, misinformation beliefs are not simply the opposite of lacking knowledge about the pandemic. It is possible to understand certain aspects of the virus and how to avoid getting infected while also having misconceptions about the pandemic. This is especially understandable during a global health crisis wherein expert knowledge itself was evolving frequently. To this end, while knowledge and misinformation beliefs correlate with each other (negatively, the more of one means less of the other), they are not the same construct and thus looking separately at how information sources relate to them is warranted. Accordingly, this section focuses on how the media that people consulted relates to misbeliefs they held about the virus.

Figure 4.8 proved helpful in revealing the relationship between number of media source types consulted and COVID-19 knowledge. To see whether a similar association of media source types and misinformation beliefs exists, figure 4.9 shows these relationships across the three countries. In the United States, there is a U-shaped relationship whereby those who consulted no sources or many are the most prone to misconceptions about the virus. Such a relationship does not hold in the other two countries. In Italy, more types of sources correlated with more misbeliefs, whereas in Switzerland there is no such relationship.

When it comes to specific information sources, the results are again different from what explains COVID-19 knowledge. As with the analyses in the previous section, the results here control for sociodemographic background and other information sources as well as political leaning. Watching news shows for pandemic information on the three big commercial networks in the United States makes no difference to holding misconceptions about the virus. Turning to the big television networks is not relevant to misinformation beliefs in Italy either. In Switzerland, however, relying on the national broadcaster SRG as a daily source for pandemic information relates to lower rates of misconceptions about the virus. The only other resource in Switzerland to link with lower levels of misbeliefs is government sites; in the United States it is relying on CNN daily; in Italy it is following sites with national statistics about virus spread.

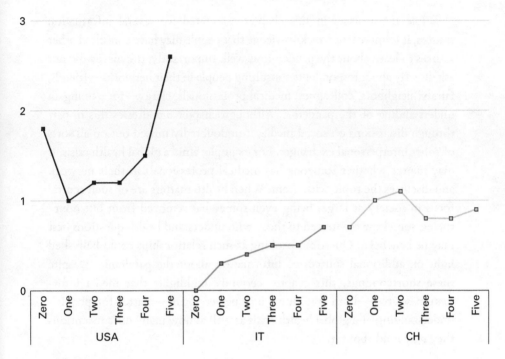

Figure 4.9
Number of misinformation beliefs held (*y*-axis) by number of media types consulted as information sources about the pandemic (*x*-axis) in the three countries.

Several sources link to increased levels of misbeliefs, such as Fox News, magazines, Facebook, Instagram, and YouTube among United States respondents. In Italy, watching local TV daily, YouTube with any frequency, and sites sharing the global spread of the virus associated with higher misconceptions. In Switzerland, listening to the radio (other than SRG) occasionally, as well as relying on Twitter, YouTube, and health sites, linked to believing more misinformation.

Results discussed in chapter 1 showed that political ideology links to misinformation beliefs in the United States and Switzerland, whereby those leaning right were more likely to hold misconceptions about the virus. This changes in both countries once we consider information sources. Once we account for where people got their information about the novel coronavirus, the variation in levels of misbeliefs by political ideology no longer holds. This suggests that media sources explain why those who lean right are more likely to hold misconceptions about COVID-19.

While the analyses in this chapter accounted for several information sources, it is important to acknowledge that people may have consulted other sources to learn about the pandemic as well. Importantly, the survey did not ask directly about respondents consulting people in their networks—friends, family, neighbors, colleagues, medical professionals, clergy—for gaining an understanding of the pandemic. Although chapter 3 addresses this in part through discussions on social media, it undoubtedly missed out on all sorts of other interpersonal exchanges. For example, amid a global health crisis, it may matter whether someone has medical professionals in their network and discusses the topic with them. When health matters are of such central focus in society at large, being even somewhat removed from but none-theless somehow connected to those who understand health questions best may be beneficial. Questions probing at such relationships could help shed light on additional sources of information about the pandemic. Despite these shortcomings, this chapter certainly highlights that media infor-mation sources—both traditional and newer forms—matter for people's understanding of a global health crisis as well as how many misconceptions they may hold about it.

This book set out to document people's digital media experiences of spring 2020 and to see whether traditional markers of social inequality reflect how they coped during these unsettled times. Inspired by the digital inequality framework, the book considered how users' sociodemographics related to their digital contexts, including their digital skills. Subsequently, the analysis showed how digital context links to variations in online behavior. To illustrate why all this matters, the exploration turned to examining how online behaviors link to knowledge and misinformation beliefs about COVID-19, finding that in many cases those operating in a more privileged digital context knew more about the pandemic.

Why might understanding the virus better and avoiding misconceptions about it matter? The hope is that it would prompt people to behave in ways that help them stay healthy, thereby relieving hospital resources already stretched thin from more burdens. It would help avoid the spread of the virus to larger parts of the population and ultimately contribute to a return to normal by reopening workplaces, schools, day-care facilities, businesses, government offices, and the many other establishments whose day-to-day operations are essential for the functioning of society. It would mean an end to the isolation that the pandemic had thrust on so many.

To show how knowledge about the pandemic mattered, I present the results of one more analysis. I have left this for the conclusion as it is the culmination of the digital inequality framework I presented in the introduction, applied specifically to the data set about people's experiences under lockdown. Does knowing more about COVID-19 translate into staying safe? Does having misconceptions about the virus reduce people's likelihood of being careful to avoid getting sick? The analyses look at how sociodemographics, digital context measures, knowledge, and misinformation beliefs related to going out for optional activities. An important recommendation from the start of the lockdown—indeed, the reason for lockdowns—was that people should stay at home and shelter in place to reduce exposure to and spread of the virus. How are digital context as well as pandemic knowledge and misconceptions related to complying with stay-at-home orders?

Figure 1.1 in chapter 1 showed the reasons for which people were leaving their homes. Among these, the following activities were in all likelihood nonessential: meeting up with friends; going out to eat; going out for beauty or personal care services; going to houses of worship; going to the movies, theater, or a concert; and going to a bar or café. Although only 4% of Italians had engaged in any of these activities, the figure was much higher in the United States and Switzerland, 23% and 22%, respectively. Did knowing more about the virus translate to people avoiding such activities? It did.

Figure 5.1 shows the differences in COVID-19 knowledge between those who left their home for optional activities (darker left columns) and those who did not (lighter right columns). In all three countries, those who were less knowledgeable about the virus were less likely to follow recommendations to shelter in place and not leave the home for nonessential activities (US: 8.5 knowledge score among those who went out vs. 9.8 among those who did not go out for optional activities; IT: 6.4 vs. 10.2; CH: 8.7 vs. 9.8).

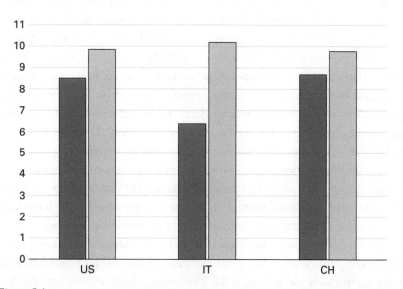

Figure 5.1
The COVID-19 knowledge score (0–11) of those who went out for optional activities (darker left columns) and those who did not (lighter right columns) across the three countries.

Having fewer misconceptions about the pandemic also related to safer behavior in the form of more limited mobility as illustrated in figure 5.2. Those who went out were more likely to believe in misinformation than those who did not (US: 2.5 misinformation beliefs among those who left the home for optional activities vs. 1.4 among those who had not left the home; IT: 2.2 vs. 0.5; CH: 1.3 vs. 0.9). In other words, understanding specifics about reducing the risk of infection and having fewer misconceptions about the same are both associated with safer behaviors.

Figure I.1 described the digital inequality framework to explain how digital context can have implications for one's life chances. I now discuss that figure in light of the empirical evidence presented in this book. Chapter 2 established that both quality of device access and digital skills are associated with people's sociodemographics background. Those from more privileged social positions have more autonomy in using the Internet and have higher digital skills. Chapters 3 and 4 showed that both sociodemographic background and digital context relate to how people use digital media. They also revealed that online behavior is linked to both knowledge and misconceptions about a deadly global pandemic. The results above show

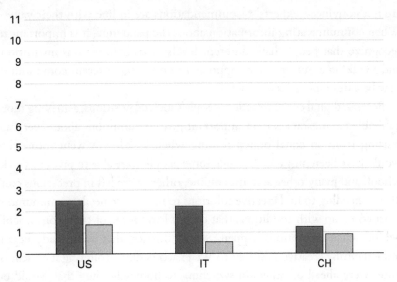

Figure 5.2
The number of misinformation beliefs about the pandemic (0–11) among those who went out for optional activities (darker left columns) and those who did not go out (lighter right columns) across the three countries.

that these then link to the real-life outcome of staying safe from a deadly virus.

Given that people's digital contexts influence their knowledge about the virus, it is significant to know that these have real-life consequences, especially because people's digital contexts vary considerably when compounded by traditional markers of inequality. When looking at racial differences in who left home for nonessential activities in the United States, results suggest that Blacks tended to do this much more than others. However, once we take knowledge about COVID-19 into account, this racial variation in behavior no longer exists. That is, African Americans who knew as much about the virus as Whites were no more likely to leave the home for nonessential activities. It is valuable to be able to identify what may explain such divergences in behavior by race. As discussed in chapter 1, African Americans have a very fraught history with the health care system in the United States due to their medical mistreatment over a long period of time. Analyzing the May 2020 US data set, my collaborator John Evans and I found that African Americans are less likely to believe that COVID-19 scientists' values align with their own.[1] The issue is not necessarily that people do not believe that scientists understand the pandemic; rather, it may be that they are not convinced experts' recommendations are in line with their values. When communicating information about the pandemic, it is important to recognize that people have different levels of trust in various institutions and so taking a multipronged approach to reaching different communities may be a desirable course of action.

Because digital contexts reflect people's societal positions, achieving more equitable circumstances is an important goal. Assumptions about everyone starting from an equal digital footing when lockdowns went into effect would have been misguided. While some people were able to pivot to work, school, and many other activities online, others were left in precarious conditions needing to find creative solutions to meet basic needs. Communities must come up with conditions that do not leave some of their constituents behind when disruptions happen. School districts that had already begun implementing creative solutions to Wi-Fi access such as offering it on school buses were ahead of others in switching to homeschooling that would be inclusive of all pupils regardless of their home environments.

This book's analyses also revealed the importance of more traditional media in unsettled times. A consistent finding is that getting information about the novel coronavirus from major television networks was linked with more pandemic knowledge. In the United States, the major commercial networks fill this role; in Switzerland and Italy, it is the public broadcast networks. At a time when there was much uncertainty and confusion, centralized sources that addressed basic facts while perhaps avoiding too many details may have worked best to convey needed health information to the population.

Another key finding is that in all three countries, people of divergent sociodemographic backgrounds select into the use of different social media platforms. If governments decide to do more communication on social media, they must attend to this diversity of information venues. For example, if agencies want to track the needs of varied populations, they must be conscious of the fact that not everybody is online and that those who are show up on different platforms at varying degrees of engagement. Accordingly, governments must continually keep abreast of this changing landscape of digital media proliferation in order to know how to calibrate and diffuse their messaging effectively and efficiently. This means regularly collecting relevant survey data about Internet use from the general population, something most governments (including of the countries studied here) do not currently do.[2]

When agencies (whether public or private) distribute forms for people to fill out, some of the constituents they are trying to reach may not hear about it and others may not know how to fill them out. If organizations are collecting measures of people's behavior through unobtrusive, automatically generated data based on specific sites and services, they must remember that such data procurement will bias against those less likely to use such services.[3] In 2010, then mayor of Newark, New Jersey (US), Cory Booker received considerable praise during a heavy snowstorm for reaching out to people on Twitter to ask who needed support with snowplowing.[4] Although it was a creative public call that certainly generated a lot of media buzz, it is unlikely that those most in need of help were using Twitter and thus would have missed this opportunity to get assistance.[5] Support services must keep people's different digital contexts in mind when attempting to meet the needs of diverse constituents.

The rollout of registrations for the COVID-19 vaccine is a vital case in point. Much of it happened through online registration. It often involved complicated forms on sites that did not always function well.[6] Those who could navigate sites swiftly could find appointments quicker than those who could not. A site might lock you out, but if you knew to try a different browser or pivot to a different device—if you had a different device to pivot to in the first place!—you may have been more successful in securing a vaccination appointment than those not aware of or not possessing such alternatives. The January 2022 rollout of the Covidtests.gov site that offered all American households free COVID-19 tests was a good example of an easy-to-use system that did not require advanced Internet skills to use.

Sometimes lack of findings from the data are interesting in their own right, which is the case with digital skills showing no relationship to misinformation beliefs about COVID-19. Much emphasis gets placed on the importance of media literacy for countering misinformation. But perhaps believing misconceptions is less about skills than some literature would have us believe. Some work has found that people share misinformation as a way to solidify certain group identities or connect with people in communities to which they belong.[7] This area is very much ripe for more research, including discovering why racial and ethnic minorities as well as disabled people were more likely to hold misconceptions about the virus.[8] These are populations that existing scholarship on misinformation has not studied at great length and may offer important and enlightening observations especially if approached from the right historical context.[9]

Another domain where this study complicated existing scholarship concerns the relationship between age and misinformation. Studies that have looked at who shares such content have tended to find that older adults are the most likely to engage in such behavior.[10] But is sharing misinformation the same as believing it? It may be that people share misinformation due to its shock value and not because they think they are spreading reliable information. Accordingly, it may be problematic to assume that certain population groups that share such content more, like older adults, do so because they are more likely to have been duped. Certainly, about the health-related content that is the focus of this book, this has not been the case. Of course, it may be that people approach health information in a crisis differently from other types of content. Whatever the reason, the findings complicate existing assumptions about the relationship of age to misinformation beliefs. Future work will need to disentangle how much of holding misconceptions is about factual misunderstandings versus ideological issues, trust in various institutions, or other possible reasons.

The goal of this book has been to bear witness to people's initial digital experiences during the COVID-19 pandemic. It has also shown that long-term inequalities that decades of research has documented concerning widespread variations in people's digital circumstances very much influenced circumstances during lockdown. In normal times one person's limited access may not be seen as something of a problem, but when everyone has to connect from home to keep abreast of a killer virus, the effects of one person's suboptimal digital condition can reverberate well beyond that individual's situation. For example, if people cannot access up-to-date information about how to stay safe, they may be putting not just themselves but other people at risk as well. If students and teachers do not have the necessary setup and know-how to connect to their classes, efforts will go toward solving logistical hurdles rather than making progress on the curriculum. If employees do not have the necessary infrastructure to connect to online meetings then not only they will get behind, but the projects dependent on them will linger as well. Given the many collective negative repercussions to lacking autonomy of use and digital skills, it behooves communities to put resources to work toward more equitable technological opportunities.

A good example of how accommodations for some can eventually become a resource for all is disabled people's efforts to make social media and other digital platforms more accessible. In fighting to have options such as video for chatting and working remotely, disabled people likely paved the way for millions of those who were scrambling at the start of the pandemic to get their home situations fit for work. As society now ponders the possibility of maintaining, even widening, these options long-term, disabled advocates should be recognized for the work they have done building the foundation of such infrastructure. One way to honor this work is to lower the hurdles they continue to face in using many technologies.

Studying people's experiences in three countries is naturally harder than doing so in one, but comes with unique contributions that a single-nation investigation cannot offer. Many findings in the book generalize across the three cases, which gives more confidence in them. At the same time, it can be valuable to recognize cross-national differences. For example, WhatsApp is relatively uncommon in the United States, whereas it has skyrocketed to the most popular social media platform status in Italy and Switzerland. Being aware of such international variations is essential when trying to generalize findings from one study to the next. It also shows the value of trying to replicate findings across borders, because they may not always generalize.

This book is based on self-reported survey measures, which are good for assessing some concepts more than others. Self-reports of online media exposure can have shortcomings because people may under or overestimate how much time they spend on different platforms.[11] Collecting media use data is possible in other ways, as has been the case for television viewing data for decades through the use of "people meters," although such methods pose their own challenges.[12] Nowadays, analyzing the automatically generated traces of people's online actions is another way to examine what people do on digital media, but such approaches have their own set of limitations.[13] Unfortunately, access to the behavior of a representative sample of users across platforms is a significant hurdle to such research. Another obstacle in the context of the present study is that such log data would be difficult to match up with many of the factors relevant to the analyses presented here, from sociodemographics to knowledge and misconceptions about COVID-19.[14] Under less pressured circumstances than a sudden global pandemic, data triangulation is preferred over relying on one data source such as a survey. Given the significant restrictions imposed by a deadly virus that put a halt to life as we knew it, having captured people's experiences at the height of initial lockdowns through one method is still preferable to not having captured them at all.

As I bring this book to a close, the United States Senate just gave bipartisan support for a $1 trillion infrastructure bill that notably includes funding for broadband. Recall the children depicted in figure 2.1 sitting outside Taco Bell using its free Wi-Fi. California has since passed Assembly Bill 156 Communications: Broadband to address such challenges. The focus in many policy discussions remains on physical access, however. While having access to high-speed connectivity is certainly a necessary condition for being truly wired in the twenty-first century, it is not sufficient. As this book has shown, access to multiple devices, and importantly, Internet skills, are important for whether and to what extent people can benefit from digital media across all areas of life, from education to employment, and importantly, self-care for staying healthy. Ensuring that people of varying backgrounds have the needed skills to use digital media effectively, efficiently, and to their benefit must be part of any policy effort that attempts to level the digital playing field, which is more and more today *the* playing field.

It may seem like 2020 was a snippet in time too unique to yield larger lessons relevant for other times. Indeed, an event that poses similar worldwide limitations may be uncommon—we can only hope!—but more geographically concentrated events that cause major disruptions are not so unusual. Continued political unrest throughout the globe and ongoing climate change will disrupt vast numbers of lives significantly. It is important to recognize the challenges people faced in the digital realm due to COVID-19 and create circumstances that mitigate these when analogous occasions arise. Historians often take great pains to understand how people of a certain era experienced events. By documenting people's behavior and sentiments in near real-time, this project sidestepped the need to come up with elaborate retroactive proxy measures and second-best markers of how digital media played a role in people's lives in this unusual period. Making sure that future disruptions to life do not replicate existing inequalities like this one did is a worthy goal to take away from understanding how digital privilege plays out in unsettled times.

ACKNOWLEDGMENTS

When this book is published, I will have been an Internet user for thirty years. I graduated from high school in Budapest in 1992, just as the Iron Curtain was disintegrating. When preparing to cross the Atlantic to pursue my college degree in the United States, my mother repeatedly emphasized to me that the moment I got to Smith College, I should ask for an email address. I may have been coming from modest financial means by American standards (after all, the Hungarian currency was not even convertible on the open exchange market then), but I was coming from an educated family with important resources in its own right. My mother, a research scientist, had been using email for years and was able to advise me about this great resource that would allow me to communicate with my parents in near real-time when international phone calls cost more than a dollar per minute. You can imagine what this meant to eighteen-year-old me as I hopped on a plane by myself to fly thousands of miles away from everyone I knew.

Over these last thirty years, I have been excited about the Internet and all the potential ways it can help people, perhaps because of how much it helped me already in those early days. But I have also been acutely aware that I had the know-how to get connected and appreciate its benefits the moment I got to college, thanks to the essential information my scientist mother had passed on to me. At that time, getting an email address was not yet automatic when arriving at college so this was a significant advantage. While I could not afford to call my family, I was nonetheless regularly in touch. This privilege was clear to me at the time, and it had me wondering in what other ways socioeconomic status might play a role in how people incorporate digital media into their lives. This question has inspired much of my research over the decades. And it inspired the specific study I decided to undertake when the world imploded in early 2020, the context of this book.

The project discussed in these pages would not have been possible without the major data-collection efforts my research team undertook in the first two months of global lockdowns in response to the COVID-19 pandemic. The indefatigable Minh Hao Nguyen and Jaelle Fuchs were with me every step of the way. Jonathan Gruber, Will Marler, Amanda Hunsaker, and Gökçe Karaoglu also contributed in important ways. Research assistants Kate Blinova, Charlotte Massiah, Alexia Röper, and in too many ways to count, Teodora Djukaric, provided valuable support during the writing process. The data cleaning, recoding, analysis, and interpretation that form the basis of this book are my work and I take full responsibility for any omissions. Team members are coauthors on resulting scholarly papers whose content is not replicated on these pages except for some brief reflections in the preface. These papers are all available at webuse .org/covid, parts of the preface draw on Hargittai et al., "From Zero to a National Data Set." Alex Maldonado provided the most amazing IT support, even from afar.

The project was not only a local effort. For their input on survey construction and in some cases translation, I thank Mark Eisenegger, John Evans, Tiziano Gerosa, Fabrizio Gilardi, Marco Gui, Monica Hamburgh, Marina Micheli, Kevin Munger, Elissa Redmiles, and Linards Udris. For assistance with methodological questions and various other odds and ends of the research process, I am grateful to Agnes Bäker, Sarah Burgard, Kerry Dobransky, Hank Farber, Jeremy Freese, Andy Guess, Ágnes Horvát, Oliver Lipps, Christine Percheski, Elissa Redmiles, and Liz Stuart. Among them are coauthors on various related publications (see webuse.org/covid) who sharpened my thinking about the material included in the book.

Lucas Freeman's keen eye for detail made for a careful edit of the full manuscript, and the book is all the better for it. At an earlier stage, my father, István Hargittai, read the full first draft and gave me valuable feedback. Will Marler and Elissa Redmiles read several chapters and shared helpful comments. I am grateful to E. Jamar for disability sensitivity editing, which, among other things, resulted in my use of the phrase "disabled people" rather than suboptimal alternatives.

Immense gratitude goes to the inspiring and tremendously supportive members of my writing accountability group: Agnes Bäker, Erin Kelly, Lisa Margulis, and Nancy Thompson. This extraordinary group ensured that despite lockdowns and isolation, the writing process was never a solitary undertaking. Thank you for lifting me up and making me smile regardless of the number of ✔s and 🐌s. This group has been nothing short of transformative for my writing and productivity over the last five years.

Additional support for helping me stay sane during the pandemic, which 100% coincided with work on this book, came from Bob Allison, Jonathan Buchbinder, Ildikó Csomó, Paul DiMaggio, Jeremy Freese, Darren Gergle, Andy Guess, Sarina Kürsteiner, Betty Leydon, John Leydon, Maggie Lu, Josh McCormick, Christine Percheski, Lauren Rivera, Shawna Samuel, Lynne Weinberg, David Weinberg, John White, Dallas Yanez, and Kati Zsinkó. I am thankful to each and every one of them. I would be remiss if I did not also mention *The Late Show with Stephen Colbert* as a constant boost to my morale during lockdowns.

Gita Manaktala at MIT Press was enthusiastic about this book project since my initial contact with her about it, even though it was a rather amorphous idea at the time, which thankfully she then helped bring into focus. Erika Barrios and Suraiya Jetha patiently answered many logistical questions and kept the production on track. Rebecca Faith and Brian Ostrander of Westchester Publishing Services helped shepherd along production. I thank the whole MIT Press team for their commitment to the book. Anonymous reviewers offered helpful feedback that pushed me to rethink what to highlight and how to present findings more clearly. I owe special thanks to Bill Dutton for suggesting "Connected in Isolation" as the title of the book. Ori Kometani deserves credit for the creative and spot-on cover.

The University of Zurich (UZH) has been an amazing environment in which to do research. I suspect many in and outside of academia have romantic notions of what it means to be a professor. I am lucky to be at an institution where these romantic notions are a reality. Extra gratitude goes to UZH's President Michael Schaepman who, as vice president of research at the time, supported the data collection with a special emergency grant. I also appreciate the University's Digital Society Initiative for providing an exciting intellectual home for the many people on campus interested in related matters. Some of the work intersected with a sabbatical, a portion of which I spent at the Studio Cascina of Villa Garbald, a marvelous university retreat space in the south of Switzerland. I thank Siska Willaert and Arnout Hostens for their hospitality during my stay. Thanks also go to Microsoft Research for its support of the survey data collection.

My parents, Magdolna and István Hargittai, have been writing books ever since I can remember, and finally I can join them in being a book author. They have supported me over the decades in ways large and small. They did the same during lockdowns through frequent video calls. So did my brother, Balázs Hargittai, who is always available for a call, whether that is to laugh or to commiserate about something. My wonderful niece Stephanie and nephew Matthew inspired lots of smiles during these unsettled times. Thanks to digital media, although thousands of miles away, my treasured family was always just a click away during this whole process, making sure we always felt connected even in isolation.

APPENDIX

SURVEY METHODOLOGY DETAILS

This section gives details about how we administered the survey, the quota sampling we applied, and what procedure I used to apply weights to the national samples. We contracted with the survey firm Cint to reach respondents who could access our survey on the Qualtrics platform using either a computer or a mobile device. Cint is an opt-in poll, and research in the past decade has shown "few or no significant differences between traditional modes [of survey administration] and opt-in online survey approaches."[1] Of course, as described in the main text, there are likely to be differences by people's digital contexts so that those with fewer devices and less reliable connections are likely not as represented.

We set quotas for age, gender, education, and region to reflect US Census figures, and age, gender, and region to ensure a diverse sample in Switzerland and Italy, respectively. Cint's respondent pool includes millions of people (15 million in the United States) gathered through a double opt-in procedure. Potential respondents are initially contacted through telephone, face-to-face interactions, email, social media, and banner ads. After potential participants fill out a form, they receive an email that requires logging into their Cint account to become part of the panel. Cint then emails potential respondents about studies and compensates them with a small remuneration for their participation.

We took care to obtain a diverse sample to make sure that different people's perspectives and experiences would be represented in the data set. The quota-sampling method mentioned above ensures that the proportion of young and old, women and men, those living in different parts of the country, and in the US case, the more and less educated, are represented at similar levels to national figures. That said, the recruitment method did not ensure that at the bivariate level (i.e., the relationship of two variables) the sample would be representative. That is, there may be more younger women or more educated men in the sample than is the case nationally. To account for this, I use weights for all country samples derived from nationally representative samples in each national context. In the case of the United States, this is the 2019 Current Population Survey;[2] in the case of Italy, the 2018 Multiscopo of the Italian National Institute of Statistics;[3] for Switzerland, the 2019 Swiss Household Panel.[4] The weight ranges are 0.51–1.70 in the United States in April, 0.49–2.36 in the United States in May, 0.42–1.60 in Italy, and 0.27–4.46 in Switzerland (with 81% falling in the 0.40–2.32 range). The weights I applied concern the national populations as a whole rather than the Internet user population of each country. While the latter would have been preferable, the baseline surveys I used do not necessarily have the needed variables to account for Internet user status.

A limitation of surveys more generally is that people may be responding mechanically without really considering the questions or their answers. Since people get financial compensation for filling out questionnaires, they

Sample descriptives (unweighted)

	United States (N = 1,374)		Italy (N = 983)		Switzerland (N = 1,350)		United States (May) (N = 1,551)	
	Mean	Percent	Mean	Percent	Mean	Percent	Mean	Percent
Age	46		50		46		47	
Income	$59,104		€60,989		CHF79,151		$59,780	
Female		54		51		50		56
Rural resident		16		13		41		19
Disabled		16		10		14		13
Higher education		29		25		39		34
Race & ethnicity								
White		65						69
Black		13						11
Asian		5						5
Native American		2						1
Hispanic		15						13

Breakdown of answers to the knowledge and misinformation questions

Multiple-choice questions	US	IT	CH
What should you do if you have come into close contact with infected people?			
Self-quarantine by staying at home as a precaution	73	79	72
Make frequent nasal washings	3	8	6
Go to the doctor and ask to get tested for the virus	19	8	15
Go out only for work or health reasons	5	5	8
What are the common symptoms of COVID-19?			
Fever and dry cough	93	97	94
Abdominal pain and cramps	1	1	2
Upset stomach and nausea	2	0	1
Headache and dizziness	3	1	3
How long does it take between catching Coronavirus and beginning to have symptoms?			
A few minutes	5	1	1
One day	9	4	6
Up to two weeks	86	93	91
Up to two months	1	1	2
Who is most at risk of serious health consequences of COVID-19?			
Older people with certain pre-existing medical conditions	95	93	88
Children, which is why schools are now closed	3	1	1
Pregnant women	1	0	2
People of Chinese descent	1	5	8
What can be said about people who have been tested positive for COVID-19 but are in good health?			
They are not contagious until they show clear symptoms	6	4	6
They are definitely going to show symptoms within a few days	8	5	6
They are contagious regardless of whether they show symptoms	82	88	81
They are already immunized and can go out in public	4	3	7

True-false questions	US	IT	CH
What are ways to reduce the risk of being infected by the Coronavirus? If you do not know, please give it your best guess. Be sure to check <u>all</u> that apply.			
Take vitamin C	36	13	20
Drink hot fluids	21	4	9
Take hot baths	16	4	4
Frequently rinse your nose with saline (salty water)	12	8	5

(continued)

Breakdown of answers to the knowledge and misinformation questions (continued)

True-false questions	US	IT	CH
Eat freshly boiled garlic	6	2	5
Clean and disinfect frequently-touched surfaces	88	90	80
Keep a distance of 6-feet with other people	90	94	94
Wash your hands often with soap	94	96	92
None of the above, there is no way to reduce risk	1	0	0

What are additional ways to reduce the risk of being infected by the Coronavirus? If you do not know, please give it your best guess. Be sure to check all that apply.

	US	IT	CH
Avoid buying products made in China	22	3	9
Avoid receiving packages from the postal service	17	4	8
Avoid physical contact with pets and other animals	16	3	9
Avoid taking anti-inflammatory drugs	13	14	21
Avoid consumption of meat products	5	2	3
Avoid consumption of dairy products	5	1	2
Avoid leaving your home	82	89	80
Avoid touching your eyes, nose and mouth with your hands	86	95	87
Avoid shaking hands with people	88	93	93
None of the above, there is no way to reduce risk	2	0	0

may prefer to skip through questions quickly to get the payment and move on rather than taking them seriously. This poses potential quality concerns. To address these, researchers can use various methods to check whether participants are paying attention. There is considerable scholarship on what methods work best, but also on how such checks can ultimately bias against certain participants.[5] Keeping all this in mind, we implemented the following attention-check measures.

First, we asked a straightforward question early in the survey. Placing this early was partly to signal to respondents that we were paying attention to whether they were paying attention. The question read as follows: "We want to make sure that people read the survey questions carefully. For this question, please mark the response that says '3'." There were six options ranging from 1–6. Later in the survey, we used a more complex approach to determine whether people were reading the questions carefully. In this case, we explained in a few sentences that the respondent was to ignore the question that was about to follow and, rather, pick the answers we just gave them to check off. This is a more complex question, but is important for knowing whether people are paying attention to questions that give some details about how to fill out the form. Since we were about to post such a query, checking for such careful attention was important. The samples used in the analyses (see "Sample descriptives" in this Appendix) exclude people who did not pass the attention check calculations.

SURVEY QUESTIONS USED IN THE ANALYSES

Included in the April 2020 surveys in the three countries.

Which of the following devices do you have available <u>at home</u> to access the Internet? Check all that apply.

- Mobile phone
- Tablet
- Laptop or desktop computer
- Smart TV
- Gaming device
- None of the above, I cannot access the Internet at home

On an average <u>weekday</u>, about how often do you use the Internet, either on a computer, tablet or phone?

- Almost constantly
- Several times a day
- About once a day
- Several times a week
- Less often

On an average <u>Saturday or Sunday</u>, about how often do you use the Internet, either on a computer, tablet or phone?

- Almost constantly
- Several times a day
- About once a day
- Several times a week [this option was only asked on the US survey]
- Less often

How familiar are you with the following computer and Internet-related items? Please choose a number between 1 and 5 where 1 represents "no understanding" and 5 represents "full understanding" of the item. [Presented as a grid.]

Answer options:
 1—None
 2
 3
 4
 5—Full

Question items:
 Advanced search
 PDF
 Spyware
 Wiki
 Cache
 Phishing

Have you left your home <u>in the past two weeks</u> for any of the following activities? Check <u>all</u> that apply.

- Buy groceries
- Shopping
- Go to the doctor or pharmacy
- Go for a walk, run or bike ride
- Walk your dog
- Go to work
- Meet up with friends
- Go out to eat
- Travel
- Beauty or personal care services
- Attend religious services
- Go to the movies, theater or a concert
- Go to a bar or cafe
- For something else not listed here, please specify:

  _____

- I have not left my home at all in the last two weeks

How familiar are you with the following computer and Internet-related items? Please choose a number between 1 and 5 where 1 represents "no understanding" and 5 represents "full understanding" of the item. [Presented as a grid.]

Answer options:
  1—None
  2
  3
  4
  5—Full

Question items:
  Privacy settings
  Meme
  Tagging
  Followers
  Viral
  Hashtag

How closely, if at all, have you been following news about the outbreak of the Coronavirus also known as COVID-19?

- Very closely
- Somewhat closely
- Not too closely
- Not at all closely

We want to make sure that people read the survey questions carefully. For this question, please mark the response that says "3".

  1
  2
  3
  4
  5
  6

What are ways to reduce the risk of being infected by the Coronavirus? If you do not know, please give it your best guess. Be sure to check <u>all</u> that apply.

- Take vitamin C
- Drink hot fluids
- Frequently rinse your nose with saline (salty water)
- Eat freshly boiled garlic
- Take hot baths
- Clean and disinfect frequently-touched surfaces
- Keep a distance of 6-feet with other people
- Wash your hands often with soap
- None of the above, there is no way to reduce risk

What are additional ways to reduce the risk of being infected by the Corona-virus? If you do not know, please give it your best guess. Be sure to check <u>all</u> that apply.

- Avoid taking anti-inflammatory drugs
- Avoid physical contact with pets and other animals
- Avoid consumption of meat products
- Avoid consumption of dairy products
- Avoid buying products made in China
- Avoid receiving packages from the postal service
- Avoid shaking hands with people
- Avoid leaving your home
- Avoid touching your eyes, nose and mouth with your hands
- None of the above, there is no way to reduce risk

Below are some questions about the Coronavirus pandemic (COVID-19). Please select the correct answer to these questions. If you don't know the correct answer, take your best guess.

What should you do if you have come into close contact with infected people?

- Self-quarantine by staying at home as a precaution
- Make frequent nasal washings
- Go to the doctor and ask to get tested for the virus
- Go out only for work or health reasons

What are the common symptoms of COVID-19?

- Fever and dry cough
- Abdominal pain and cramps
- Upset stomach and nausea
- Headache and dizziness

How long does it take between catching Coronavirus and beginning to have symptoms?

- A few minutes
- One day
- Up to two weeks
- Up to two months

Who is most at risk of serious health consequences of COVID-19?

- Older people with certain pre-existing medical conditions
- Children, which is why schools are now closed
- Pregnant women
- People of Chinese descent

What can be said about people who have been tested positive for COVID-19 but are in good health?

- They are not contagious until they show clear symptoms
- They are definitely going to show symptoms within a few days
- They are contagious regardless of whether they show symptoms
- They are already immunized and can go out in public

We are interested in how people get information about the Coronavirus pandemic (COVID-19). How often have you gotten information about Coronavirus from the following sources? Be sure to select the source whether you saw the information from it online or offline. [Presented as a grid.]

Answer options:
   Never
   Few times a week or less
   Daily or almost daily

Question items (US):
   News show on ABC, CBS, or NBC
   Fox News
   MSNBC
   CNN
   Other cable news channels
   PBS
   NPR (National Public Radio)
   Other radio
   National newspaper
   Local newspaper
   Magazine
   Late night talk show

Question items (IT):
   News show on national TV stations (Rai, Mediaset, La7, Sky)
   News shows on local TV stations
   Radio
   National newspaper
   Local newspaper
   Magazine

Question items (CH):
   SRG-SSR
   Private TV from abroad
   Public TV from abroad
   Private Swiss TV
   Private Swiss radio
   Subscription newspaper
   Free newspaper
   Magazine

How often have the following been sources of information for you regarding the Coronavirus pandemic? [Presented as a grid.]

Answer options:
   Never
   Few times a week or less
   Daily or almost daily

Question items:
  Facebook
  Twitter
  Instagram
  WhatsApp [included on IT & CH versions only]
  YouTube (excluding videos from mainstream news outlets)
  Online-only news media
  Government websites
  Health websites
  Websites displaying numbers about the global spread
  Websites displaying numbers about its spread in [name of country]

How often, if ever, do you use the following sites and services? [Presented as a grid.]

Answer options:
  Never
  Few times a week or less
  Daily or almost daily

Question items:
  Facebook
  Instagram
  Twitter
  Snapchat
  TikTok
  Reddit
  WhatsApp
  Google or another search engine

Have you seen the following types of information about Coronavirus on the social media platforms listed below? [Presented as a grid.]

Answer options:
  on Facebook
  on Instagram
  on WhatsApp [included on IT & CH versions only]
  on Twitter
  I have not seen such content

Question items:
  Seen tips on how to avoid getting infected
  Seen information about symptoms
  Seen numbers or charts about its spread
  Seen government rules about what people are allowed to do
  Seen religious sentiments and teachings related to it
  Seen humor, jokes, funny content related to it
  Seen gratitude expressed toward health care workers
  Seen news coverage about it

Have you **shared** the following types of information about Coronavirus on
the social media platforms listed below? [Presented as a grid.]

Answer options:
  on Facebook
  on Instagram
  on WhatsApp [included on IT & CH versions only]
  on Twitter
  I have not shared such content

Question items:
  Shared tips on how to avoid getting infected
  Shared information about symptoms
  Shared numbers or charts about its spread
  Shared government rules about what people are allowed to do
  Shared religious sentiments and teachings related to it
  Shared humor, jokes, funny content related to it
  Shared gratitude expressed toward health care workers
  Shared news coverage about it

Have you had any of the following types of interactions about the Coronavirus
pandemic on the social media platforms listed below? [Presented as a grid.]

Answer options:
  on Facebook
  on Instagram
  on WhatsApp [included on IT & CH versions only]
  on Twitter
  This has not happened

Question items:
   Asked a question about it
   Received an answer to a question you asked about it
   Answered someone else's question about it
   Participated in a discussion about it
   Posted about your own experiences related to it
   Asked for support
   Offered support
   Received support
   Corrected someone else's post or comment about it

Below are some institutions in this country. As far as the people running these institutions are concerned, would you say you have a great deal of confidence, only some confidence, or hardly any confidence at all in them in <u>addressing the Coronavirus pandemic</u>? [Presented as a grid.]

Answer options:
   A great deal of confidence
   Only some confidence
   Hardly any confidence

Question items:
   Medical system
   The federal government
   Local/state government
   Business leaders
   Religious leaders

On a scale of 1–5 where 1 means "Very well" and 5 means "Not at all," how well do the following groups understand the spread and health impacts of Coronavirus (COVID-19)? [Presented as a grid.]

Answer options:
   1—Very well
   2
   3
   4
   5—Not at all

Question items:
 Medical researchers/Doctors
 Federal government political leaders
 State government political leaders
 Business leaders
 Journalists
 Religious leaders

How does the current Coronavirus pandemic affect your circumstances at home? Check all that apply.

- More calmness and relaxation
- Strengthened family/partnership
- More personal time
- Tensions and conflicts in the household
- Feeling trapped
- Excessive childcare demands
- Lack of personal space or alone time
- Time spent on homeschooling
- I have not experienced any of these

Since the Coronavirus outbreak, have any of the following been worrying you more than usual, even if only in a minor way? Check all that apply.

- Your pet
- Finances
- Getting food or medication
- Your own safety / security
- Internet access
- Boredom
- Future plans
- None of these

Since the Coronavirus pandemic, how often have you used the following methods to communicate with friends and family who do not live in your household? Do not include work-related communication. [Presented as a grid.]

Answer options:
  Daily/almost daily
  Few times a week
  Less than weekly
  Never

Question items:
  Voice calls
  Video calls
  Text messages (using any messaging app)
  Email
  Social media
  Online games
  Postal mail

Compared to before the Coronavirus pandemic, has your communication with friends and family who do not live in your household increased, decreased or remained the same for these methods? Do not include work-related communication.

Answer options:
  More
  About the same
  Less

Question items:
  Voice calls
  Video calls
  Text messages (using any messaging app)
  Email
  Social media
  Online games
  Postal mail

Below are some statements about how people might feel. When thinking about the past two weeks, please indicate how much you disagree or agree with the statements.

There are no right or wrong answers. Try to not spend too much time on any one statement but give the answer which seems to describe your feelings best. [Presented as a grid.]

Answer options:
  Strongly disagree
  Disagree
  Mildly disagree
  Mildly agree
  Agree
  Strongly agree

Question items:
  I felt disconnected from the world around me
  Even around people I know, I didn't feel that I really belonged
  I felt so distant from people
  I had no sense of togetherness with my peers
  I caught myself losing all sense of connectedness with society
  I didn't feel I participated with anyone or any group

We would like to get a sense of your general preferences.

Preferences can depend on different factors. In this question, we have a specific request. To show that you have read this much, just go ahead and select both red and green among the options below, no matter what your favorite color is. Yes, ignore the question below and select both of those options.

What is your favorite color?

- White
- Black
- Red
- Pink
- Green
- Blue

Have you been diagnosed with the Coronavirus (COVID-19)?

- Yes
- No

Do you know any people who have been diagnosed with Coronavirus (COVID-19)?

- Yes
- No

What is your relationship to this person? If you know more than one person who has been diagnosed with Coronavirus (COVID-19), please mark all relationships that apply.

- Family
- Friend
- Co-worker
- Professional contact
- Neighbor
- Acquaintance
- Other, please specify:

_____

Has anyone you know died from the disease?

- Yes
- No

We are sorry for your loss.

Do you have any of the following medical conditions? Check all that apply.

- High blood pressure
- Diabetes
- Cardiovascular disease, heart disease
- Chronic respiratory disease, lung disease (such as asthma, COPD)
- Cancer
- Conditions and therapies that weaken the immune system
- Clinically-diagnosed depression
- Clinically-diagnosed anxiety
- Another clinically-diagnosed mental health problem
- I take immunosuppressive medication
- I am pregnant
- None of the above

Do you have any of the following long-lasting conditions? Check all that apply.

- Blindness or severe vision impairment even with glasses or contact lenses
- Deafness or a severe hearing impairment even with a hearing aid
- Serious difficulty having your speech understood
- Serious difficulty walking or climbing stairs
- Serious difficulty dressing or bathing
- Serious difficulty typing on a traditional computer keyboard
- Serious difficulty concentrating, remembering, or making decisions
- Serious difficulty going outside the house alone
- None of the above

In what year were you born? (four digits please)

_____

Are you:

- Male
- Female
- Other, please specify:

_____

Are you of Hispanic or Latino descent?

- No
- Yes

Please check one or more categories below to indicate what race or races you consider yourself to be.

- White
- Black/African American
- Asian
- American Indian or Alaska Native
- Native Hawaiian or Pacific Islander
- Other, please specify:

_____

What is the highest level of school you have completed or the highest degree you have received?

- Less than high school degree
- High school graduate (high school diploma or equivalent including GED)
- Some college but no degree
- Associate's degree
- Bachelor's degree
- Advanced degree (e.g., Master's, doctorate)

How many adults currently live in your household, including you?

- 1
- 2
- 3
- 4
- 5 or more

Are you currently married or in a committed romantic relationship?

- Yes
- No

Are you currently living in the same household with your romantic partner/ spouse?

- Yes
- No

How many children under the age of 18 currently live in your household, including those who live there part time?

None
- 1
- 2
- 3
- 4
- 5 or more

Who is currently providing childcare <u>in your home</u>? Check <u>all</u> that apply.

- I am
- My spouse/partner
- Other family member
- Friend
- Neighbor
- Paid childcare provider—in person
- Paid childcare provider—virtually

Do you currently have any childcare <u>outside</u> of the home?

- No
- Yes, up to 10 hours a week
- Yes, 10–20 hours a week
- Yes, more than 20 hours a week

Have you had a change in your employment status due to the Coronavirus?

- Yes
- No

What change has happened in your employment status? Check all that apply.

- Got laid off
- On forced unpaid leave
- On voluntary unpaid leave
- Working reduced hours
- Working increased hours
- Got new job that is better than earlier job
- Got new job that is worse than earlier job
- Got new job that is about the same as earlier job
- Something else, please explain:

_____

Generally speaking, do you usually think of yourself as a Republican, a Democrat, an Independent, or what? [US] [Asked as a political orientation question on 0–10-point scale in Italy and Switzerland where 0 was left and 10 was right. US questions recoded into 1–6 scale.]

- Republican
- Democrat
- Independent
- No preference

Would you call yourself a strong Republican or a not very strong? [US]

- Strong
- Not very strong

Would you call yourself a strong Democrat or a not very strong? [US]

- Strong
- Not very strong

Do you think of yourself as closer to the Republican or Democratic party? [US]

- Republican
- Democratic

How would you describe the type of community you live in?

- A big city
- The suburbs or outskirts of a big city
- A town or a small city
- A rural area

In which state do you currently reside? [US] [Asked about regions in Italy and cantons in Switzerland.]

- [US states]

Which one of the following includes your total HOUSEHOLD income for last year, before taxes? [US] [Asked in local currency and adjusted ranges in Italy and Switzerland.]

- Less than $10,000
- $10,000 to under $20,000
- $20,000 to under $30,000

- $30,000 to under $40,000
- $40,000 to under $50,000
- $50,000 to under $65,000
- $65,000 to under $80,000
- $80,000 to under $100,000
- $100,000 to under $125,000
- $125,000 to under $150,000
- $150,000 to under $200,000
- $200,000 to under $250,000
- $250,000 or more

Included on the May 2020 survey in the United States.

During the Coronavirus pandemic, some people have needed help with some activities. Have you needed help with each of the following? For each, indicate whether you needed help and if yes, whether you received the help you needed. [Presented as a grid.]

Answer options:
> Have not needed such help
> Have needed such help, but could not get it
> Have needed such help and got it

Question items:
> Access to masks
> Getting groceries or picking up medications
> Keeping in regular contact with others while being unable to leave the home or receive visitors
> Help with childcare
> How to do something on smartphone, tablet or computer
> How to find or understand information related to COVID-19
> How to apply for federal assistance related to COVID-19
> Financial assistance to cover everyday expenses
> Needed support in another way, please specify:

_____

People sometimes have questions about or run into problems with using technology such as downloading apps, using video services, learning new programs, and sharing content. During the Coronavirus pandemic, have you had questions about or problems with any of the following? Check <u>all</u> that apply.

- Finding an app or program you need
- Downloading/installing/running an app or program
- Signing up for a new online service
- Getting video chat to work including problems with the microphone, sound, etc
- Sharing content online
- Getting the Internet to work on a device
- Setting up a new device
- Finding people you want to contact
- Something else, please specify:

  _____

- None, I have had no questions about or problems with technology during the pandemic

Generally speaking, how often have you been able to find a solution to your questions about and problems with technology during the Coronavirus pandemic?

- None of the time
- Some of the time
- About half of the time
- Much of the time
- All of the time

NOTES

PREFACE

1. Albright, "On Unexpected Events"; Tufekci, "How Social Media Took Us."

2. Hargittai et al., "From Zero to a National Data Set."

INTRODUCTION

1. DiMaggio and Hargittai, "From the 'Digital Divide'."

2. Roser, Ritchie, and Ortiz-Ospina, "Internet."

3. Ono and Mori, "COVID-19 and Telework."

4. Chen and Wellman, "Charting Digital Divides"; Correa, Pavez, and Contreras, "Digital Inclusion through Mobile Phones?"; Schradie, "Digital Production Gap"; Robinson et al., "Digital Inequalities"; Warschauer, *Technology and Social Inclusion*; Mossberger, Tolbert, and Stansbury, *Virtual Inequality*; Norris, *Digital Divide*; Ono and Zavodny, "Digital Inequality"; Livingstone and Helsper, "Gradations in Digital Inclusion."

5. Choi and Stvilia, "Web Credibility Assessment"; van Deursen, "Internet Skill-Related Problems"; Hargittai, "Second-Level Digital Divide."

6. Motta, Stecula, and Farhart, "How Right-Leaning Media Coverage"; Chipidza et al., "Topic Analysis."

7. National Telecommunications and Information Administration, "Falling Through the Net."

8. DiMaggio and Hargittai, "From the 'Digital Divide'," 2.

9. Hoffman, Novak, and Schlosser, "Evolution of the Digital Divide"; Katz, Rice, and Aspden, "The Internet, 1995–2000."

10. Attewell, "First and Second Digital Divides"; Fink and Kenny, "W(h)Ither the Digital Divide?"; Hargittai, "Second-Level Digital Divide"; Katz and Rice, *Social Consequences of Internet Use*; Mossberger, Tolbert, and Stansbury, *Virtual Inequality*; Natriello, "Bridging the Second Digital Divide"; Selwyn, "Defining the 'Digital Divide'"; van Dijk, "Widening Information Gaps"; Warschauer, *Technology and Social Inclusion*.

11. DiMaggio and Hargittai, "From the 'Digital Divide'."

12. Hargittai and Micheli, "Internet Skills and Why They Matter."

13. Acquisti and Gross, "Imagined Communities"; Marwick and boyd, "Networked Privacy"; Tufekci, "Can You See Me Now?"; Tufekci, "Grooming, Gossip, Facebook and Myspace."

14. Milne, Labrecque, and Cromer, "Toward an Understanding"; Smit, Van Noort, and Voorveld, "Understanding Online Behavioural Advertising"; Büchi et al., "Digital Inequalities."

15. Sunstein, "Deciding by Default."

16. Gui and Büchi, "From Use to Overuse"; Nguyen et al., "Trading Spaces."

17. Goffman, *The Presentation of Self*.

18. Hsieh, "Online Social Networking Skills."

19. Choi and Stvilia, "Web Credibility Assessment"; Metzger, "Making Sense of Credibility."

20. Hargittai, "Second-Level Digital Divide"; van Deursen, "Internet Skills: Vital Assets."

21. Litt et al., "Awkward Encounters"; Gui and Büchi, "From Use to Overuse"; Litt and Hargittai, "Just Cast the Net"; boyd and Hargittai, "Facebook Privacy Settings"; Vitak, "Impact of Context Collapse."

22. Hargittai and Hinnant, "Digital Inequality"; van Deursen, van Dijk, and ten Klooster, "Increasing Inequalities"; Ingen and Matzat, "Inequality in Mobilizing Online Help."

23. Arora, *Next Billion Users*.

24. Redmiles and Buntain, "How Feelings of Trust"; Redmiles, Kross, and Mazurek, "Where Is the Digital Divide?"; Vitak et al., "Too Good to Be True."

25. Button et al., "Online Frauds"; Marwick, "Why Do People Share Fake News?"; Burnes et al., "Prevalence of Financial Fraud"; Gangadharan, "The Downside of Digital Inclusion"; Gui and Büchi, "From Use to Overuse."

26. Hayes and Lawless, "The Decline of Local News."

27. OECD, "Crisis Squeezes Income."

28. Pew Research Center, "Internet/Broadband Fact Sheet."

29. World Bank, "Individuals Using the Internet."

30. König, Seifert, and Doh, "Internet Use among Older Europeans."

31. World Bank, "Individuals Using the Internet."

32. Kappeler, Festic, and Latzer, "Who Remains Offline and Why?"

33. Lazer et al., "Measuring Human Behavior."

34. Lazer et al.

35. Hargittai, *Handbook of Digital Inequality*.

36. Appiah, "Case for Capitalizing the 'B'"; National Association of Black Journalists, "NABJ Style Guide."

CHAPTER 1

1. World Health Organization, "Advice for the Public."

2. Brzezinski et al., "The Covid-19 Pandemic."

3. Vagnoni, "Researchers Find Coronavirus Was Circulating."

4. Lawler, "Timeline."

5. Kelén, Fulterer, and Skinner, "Coronavirus."

6. Steenhuysen and Whitcomb, "Washington State Man."

7. Wikipedia, "Template:COVID-19 Pandemic Data."

8. Igielnik, "Rising Share of Working Parents."

9. Elks, "Chores and Childcare"; Sevilla and Smith, "Baby Steps."

10. Evon, "'Sacrifice the Weak' Sign?"

11. Bay City News, "Antioch City Council Removes Planning Commissioner."

12. Shapiro, "People With Disabilities"; Shapiro, "Oregon Hospitals Didn't Have Shortages."

13. Hoskin and Finch, "Disabled People."

14. Wong, "Does the Coronavirus Pandemic?"

15. "2010 ADA Regulations."

16. Gram, "For Many Caregivers."

17. Ministero dell'Economia e delle Finanze, "I provvedimenti del Governo."

18. Swiss Confederation. "Short-Time Working."

19. Alsan and Wanamaker, "Tuskegee"; Gamble, "Under the Shadow of Tuskegee."

20. Sullivan, "Trust, Risk, and Race."

21. World Health Organization, "Advice for the Public."

22. DeWalt and Pignone, "Reading Is Fundamental."

23. Schwei et al., "Impact of Sociodemographic Factors."

24. Mitchell et al., "How Americans Navigated the News"; Barrios and Hochberg, "Risk Perceptions and Politics"; Pew Research Center, "Republicans, Democrats."

25. Mitchell et al., "How Americans Navigated the News."

26. World Health Organization, "Advice for the Public."

27. Center for the Study of Hate & Extremism, "Report to the Nation"; Allam, "FBI Report"; Daube, "Backlash."

28. Guess et al., "Cracking Open the News Feed."

29. Palca, "NIH Panel Recommends."

CHAPTER 2

1. Kang, "Parking Lots."

2. Goldberg, "Digital Divide."

3. Stelitano et al., "The Digital Divide and COVID-19."

4. Rogers and Ellerson Ng, "Report of Initial Findings."

5. Ebrahimji, "School Sends California Family a Hotspot."

6. Koumpilova, "Chicago Is Spending $50M."

7. Koumpilova, "Chicago Helped 55,000 Students."

8. Hannon, "School Bus Wi-Fi Hotspots."

9. Global Strategy Group, "New America Higher Ed Survey."

10. Fishman and Hiler, "New Polling from New America."

11. Anderson, "Accessibility Suffers During Pandemic."

12. Kang, "Parking Lots."

13. CBSLA Staff, "LA County Libraries," 19.

14. Kang, "Parking Lots."

15. Stern, "New FCC Chairman."

16. Razzi, "La tecnologia è un bene comune."

17. il Fatto Nisseno Editorial Board, "Coronavirus, Orlando to Conte.'"

18. Ferraglioni, "Dopo un anno di Dad."

19. Mascheroni et al., "La didattica a distanza durante."

20. Rai Scuola, "La Scuola in tivù."

21. Kessler, "Coronavirus deckt Mängel auf"; Häberli, "Coronavirus."

22. Huber et al., *COVID-19.*

23. DiMaggio and Hargittai, "From the 'Digital Divide'."

24. Dutton and Blank, "The Emergence of next Generation Internet Users."

25. DiMaggio and Hargittai, "From the 'Digital Divide'"; Hassani, "Locating Digital Divides."

26. Gonzales, "The Contemporary US Digital Divide"; Gonzales, McCrory Calarco, and Lynch, "Technology Problems and Student Achievement Gaps."

27. Dutton and Blank, "Emergence."

28. Gonzales, "Contemporary US Digital Divide."

29. Dutton and Blank, "Emergence."

30. Hargittai and Hinnant, "Digital Inequality."

31. Blank and Lutz, "Benefits and Harms from Internet Use"; Hargittai and Hinnant, "Digital Inequality"; Hargittai, "Digital Reproduction of Inequality."

32. Hargittai and Hsieh, "Succinct Survey Measures."

33. Hargittai, "Survey Measures."

34. European Commission, "Digital Economy."

35. Hargittai and Shaw, "Comparing Internet Experiences."

36. Vaidhyanathan, "Generational Myth."

37. Prensky, "Digital Natives, Digital Immigrants."

38. Quan-Haase et al., "Dividing the Grey Divide"; Hargittai, Piper, and Morris, "From Internet Access to Internet Skills"; Hargittai and Dobransky, "Old Dogs, New Clicks"; Hunsaker et al., "Unsung Helpers."

39. Micheli, Redmiles, and Hargittai, "Help Wanted."

40. Hunsaker et al., "Unsung Helpers."

41. Dobransky and Hargittai, "Disability Divide."

42. Vicente and López, "Multidimensional Analysis."

43. Hargittai and Shafer, "Differences."

44. Correll, "Gender," 1691, emphasis added.

45. Hargittai and Walejko, "Participation Divide."

46. Alper, *Giving Voice*, and Ellcessor, *Restricted Access*, are notable exceptions.

47. Dobransky and Hargittai, "Disability Divide"; Dobransky and Hargittai, "Closing Skills Gap"; Goggin, "Disability, Internet, and Digital Inequality."

48. Duplaga, "Digital Divide"; Vicente and López, "Multidimensional Analysis."

49. Lazar and Jaeger, "Reducing Barriers to Online Access."

50. US Government Publishing Office, US Rehabilitation Act of 1973.

51. Goggin, "Disability and Mobile Internet."

52. European Commission, "European Accessibility Act."

53. Schlomann et al., "Assistive Technology "; Lancioni et al., "Speech-Generating Devices"; Doukas et al., "Digital Cities of the Future."

54. Meiselwitz, Wentz, and Lazar, "Universal Usability."

55. Dobransky and Hargittai, "Unrealized Potential."

56. Duplaga, "Digital Divide."

CHAPTER 3

1. Treem et al., "What We Are Talking About."

2. Smith and Kollock, eds., *Communities in Cyberspace*.

3. boyd and Ellison, "Social Network Sites"; Treem and Leonardi, "Social Media Use in Organizations."

4. Weinstein, "Social Media See-Saw"; Micheli, "Social Networking Sites"; Viner et al., "Roles of Cyberbullying,"; Sharma et al., "Zika Virus Pandemic"; McLaughlin and Vitak, "Norm Evolution"; Wohn and LaRose, "Effects of Loneliness."

5. Gerson, Plagnol, and Corr, "Subjective Well-Being"; Michikyan, Subrahmanyam, and Dennis, "Can You Tell Who I Am?"; Brown and Tiggemann, "Attractive Celebrity"; Tiggemann and Anderberg, "Social Media Is Not Real."

6. Shaw and Hargittai, "Pipeline of Online Participation Inequalities."

7. van Dijck, *Culture of Connectivity*; Hargittai, "Potential Biases in Big Data."

8. Hargittai, "Making of a Popular Photo App."

9. Light, Burgess, and Duguay, "Walkthrough Method."

10. boyd, "White Flight in Networked Publics?"; Hargittai, "Whose Space?"; Hargittai, "Is Bigger Always Better?"; Blank and Lutz, "Representativeness of Social Media"; Sheldon and Bryant, "Instagram"; Mellon and Prosser, "Twitter and Facebook"; Hellemans, Willems, and Brengman, "Daily Active Users"; Gazit, Aharony, and Amichai-Hamburger, "Tell Me Who You Are."

11. Hargittai and Litt, "Tweet Smell of Celebrity Success."

12. Pew Research Center, "Demographics of Social Media Users."

13. boyd and Heer, "Profiles as Conversation"; Marwick, "To Catch a Predator?"

14. Jackson, Bailey, and Welles, *#HashtagActivism*.

15. Hargittai, "Is Bigger Always Better?"

16. Guess, Nagler, and Tucker, "Less than You Think."

17. Boudreau, "Social Media Accessibility."

18. Hollier, "Growing Importance"; Kent, "Self-Tracking Health Over Time."

19. Holton, "Facebook Accessibility."

CHAPTER 4

1. World Health Organization, "1st WHO Infodemiology Conference."

2. Tichenor, Donohue, and Olien, "Mass Media Flow."

3. Gaziano, "Knowledge Gap."

4. Hwang and Jeong, "Revisiting the Knowledge Gap Hypothesis"; Lind and Boomgaarden, "What We Do and Don't Know"; Viswanath and Finnegan, "Knowledge Gap Hypothesis."

5. Morrow, "*Sesame Street.*"

6. "Television History—The First 75 Years."

7. Ball and Bogatz, *Summary of the Major Findings.*

8. McQuail, *Audience Analysis.*

9. Eveland and Scheufele, "Connecting News Media Use."

10. Jenssen, "Widening or Closing the Knowledge Gap?"

11. Lind and Boomgaarden, "What We Do and Don't Know."

12. Cacciatore, Scheufele, and Corley, "Another (Methodological) Look"; Jeffres, Neuendorf, and Atkin, "Acquiring Knowledge."

13. Bonfadelli, "Internet and Knowledge Gaps"; Hargittai and Hinnant, "Digital Inequality"; van Deursen, van Dijk, and ten Klooster, "Increasing Inequalities"; Zillien and Hargittai, "Digital Distinction."

14. Hofer et al., "Older Adults' Online Information Seeking."

15. Boukes and Vliegenthart, "Knowledge Gap Hypothesis Across Modality."

16. Beckers et al., "What Do People Learn?"

17. Morris and Morris, "Evolving Learning."

18. Karpf, "Social Science Research Methods."

19. Geiger, "Key Findings"; Fletcher and Nielsen, "Are People Incidentally Exposed?"

20. Bonfadelli, "Internet and Knowledge Gaps."

21. Hermida et al., "Share, Like, Recommend."

22. Yoo and de Zuniga, "Connecting Blog, Twitter, and Facebook Use."

23. Hwang and Jeong, "Revisiting the Knowledge Gap Hypothesis."

24. Cacciatore, Scheufele, and Corley, "Another (Methodological) Look."

25. Bonfadelli, "Mass Media and Biotechnology."

26. Spence, Lachlan, and Burke, "Differences in Crisis Knowledge."

27. Ho, "Knowledge Gap Hypothesis in Singapore."

28. Deurenberg-Yap et al., "Singaporean Response."

29. McGonagle, "'Fake News'"; Guess and Lyons, "Misinformation, Disinformation, and Online Propaganda"; Gilardi, *Digital Technology, Politics, and Policy-Making*.

30. Jurkowitz and Mitchell, "Americans."

31. Allcott and Gentzkow, "Social Media and Fake News"; Guess et al., "Cracking Open the News Feed."

32. Dutton and Fernandez, "How Susceptible Are Internet Users?"

33. Hargittai, "Second-Level Digital Divide"; van Deursen and van Dijk, "Internet Skills Performance Tests."

34. Shaw and Hargittai, "Pipeline of Online Participation Inequalities."

35. Gui and Büchi, "From Use to Overuse."

36. Mitchell, Oliphant, and Shearer, "About Seven-in-Ten."

37. For example, Brewer and Cao, "Candidate Appearances on Soft News."

CONCLUSION

1. Evans and Hargittai, "Who Doesn't Trust Fauci?"

2. Hargittai, "Broadband Subsidies Important."

3. Hargittai, "Potential Biases in Big Data"; Lazer et al., "Measuring Human Behavior."

4. Gregory, "Cory Booker."

5. Hargittai, "Potential Biases in Big Data."

6. Wetsman, "Older Adults."

7. Marwick, "Why Do People Share Fake News?"

8. Prasad, "Anti-Science Misinformation and Conspiracies."

9. Jaiswal, LoSchiavo, and Perlman, "Disinformation, Misinformation."

10. Guess, Nagler, and Tucker, "Less than You Think."

11. Guess et al., "How Accurate Are Survey Responses?"

12. Webster and Wakshlag, "Measuring Exposure to Television"; Klopfenstein, "Audience Measurement."

13. Karpf, "Social Science Research Methods."

14. Lazer et al., "Measuring Human Behavior."

APPENDIX

1. Ansolabehere and Schaffner, "Taking the Study of Political Behavior Online," 89.

2. United States Census Bureau, "Current Population Survey (CPS)."

3. Italian National Institute of Statistics, "Multiscopo ISTAT—Aspects of Daily Life (2016)."

4. Tillmann et al., "The Swiss Household Panel Study."

5. Berinsky, Margolis, and Sances, "Separating the Shirkers from the Workers?"

BIBLIOGRAPHY

Acquisti, A., and R. Gross. "Imagined Communities: Awareness, Information Sharing, and Privacy on the Facebook." In *Privacy Enhancing Technologies*, edited by P. Golle and G. Danezis, 36–58. Cambridge, UK: Robinson College, 2006.

Albright, Karen. "On Unexpected Events: Navigating the Sudden Research Opportunity of 9/11." In *Research Confidential: Solutions to Problems Most Social Scientists Pretend They Never Have*, edited by Eszter Hargittai. Ann Arbor, MI: University of Michigan Press, 2009.

Allam, Hannah. "FBI Report: Bias-Motivated Killings at Record High Amid Nationwide Rise in Hate Crime." *NPR.org*, November 16, 2020. https://www.npr.org/2020/11/16/935439777/fbi-report-bias-motivated-killings-at-record-high-amid-nationwide-rise-in-hate-c.

Allcott, Hunt, and Matthew Gentzkow. "Social Media and Fake News in the 2016 Election." *Journal of Economic Perspectives* 31, no. 2 (May 2017): 211–236. https://doi.org/10.1257/jep.31.2.211.

Alper, Meryl. *Giving Voice: Mobile Communication, Disability, and Inequality*. The John D. and Catherine T. MacArthur Foundation Series on Digital Media and Learning. Cambridge, MA: MIT Press, 2017.

Alsan, Marcella, and Marianne Wanamaker. "Tuskegee and the Health of Black Men." *The Quarterly Journal of Economics* 133, no. 1 (February 2018): 407–455. https://doi.org/10.1093/qje/qjx029.

Anderson, Greta. "Accessibility Suffers during Pandemic." *Inside Higher Ed*, August 6, 2020. https://www.insidehighered.com/news/2020/04/06/remote-learning-shift-leaves-students-disabilities-behind.

Ansolabehere, Setphen, and Brian F. Schaffner. "Taking the Study of Political Behavior Online." In *The Oxford Handbook of Polling and Survey Methods*, edited by Lonna Rae Atkeson and R. Michael Alvarez, 76–96. Oxford: Oxford University Press, 2018. https://doi.org/10.1093/oxfordhb/9780190213299.013.6.

Appiah, Kwame Anthony. "The Case for Capitalizing the *B* in Black." *The Atlantic*, June 18, 2020. https://www.theatlantic.com/ideas/archive/2020/06/time-to-capitalize-blackand-white/613159/.

Arora, Payal. *The Next Billion Users: Digital Life Beyond the West*. Cambridge, MA: Harvard University Press, 2019.

Attewell, P. "The First and Second Digital Divides." *Sociology of Education* 74, no. 3 (2001): 252–259.

Ball, Samuel, and Gerry Ann Bogatz. *A Summary of the Major Findings in "The First Year of Sesame Street: An Evaluation."* Princeton, NJ: Educational Testing Service, 1970. https://eric.ed.gov/?id=ED122799.

Barrios, John M., and Yael V. Hochberg. "Risk Perceptions and Politics: Evidence from the COVID-19 Pandemic." *Journal of Financial Economics*, 142, no. 2 (2021): 862–879. https://doi.org/10.1016/j.jfineco.2021.05.039.

Bay City News. "Antioch City Council Removes Planning Commissioner after Viral Uproar over Remarks." *NBC Bay Area*, May 2, 2020. https://www.nbcbayarea.com/news/coronavirus/antioch-city-council-removes-planning-commissioner-after-viral-uproar-over-remarks/2283277/.

Beckers, Kathleen, Peter Van Aelst, Pascal Verhoest, and Leen d'Haenens. "What Do People Learn from Following the News? A Diary Study on the Influence of Media Use on Knowledge of Current News Stories." *European Journal of Communication*, 36, no. 3 (2021): 254–269. https://doi.org/10.1177/0267323120978724.

Berinsky, Adam J., Michele F. Margolis, and Michael W. Sances. "Separating the Shirkers from the Workers? Making Sure Respondents Pay Attention on Self-Administered Surveys." *American Journal of Political Science* 58, no. 3 (2014): 739–753. https://doi.org/10.1111/ajps.12081.

Blank, Grant, and Christoph Lutz. "Benefits and Harms from Internet Use: A Differentiated Analysis of Great Britain." *New Media & Society* 20, no. 2 (2018): 618–640. https://doi.org/10.1177/1461444816667135.

Blank, Grant, and Christoph Lutz. "Representativeness of Social Media in Great Britain: Investigating Facebook, LinkedIn, Twitter, Pinterest, Google+, and Instagram." *American Behavioral Scientist* 61, no. 7 (2017): 741–756. https://doi.org/10.1177/0002764217717559.

Bonfadelli, Heinz. "The Internet and Knowledge Gaps: A Theoretical and Empirical Investigation." *European Journal of Communication* 17, no. 1 (2002): 65–84. https://doi.org/10.1177/0267323102017001607.

Bonfadelli, Heinz. "Mass Media and Biotechnology: Knowledge Gaps Within and Between European Countries." *International Journal of Public Opinion Research* 17, no. 1 (2005): 42–62. https://doi.org/10.1093/ijpor/edh056.

Boudreau, Denis. "Social Media Accessibility: Where Are We Today: A Modest Attempt at Awakening the Giants." Presented at the CSUN 2012, San Diego, CA, March 1, 2012. https://www.slideshare.net/AccessibiliteWeb/20120301-web041socialmedia.

Boukes, Mark, and Rens Vliegenthart. "The Knowledge Gap Hypothesis Across Modality: Differential Acquisition of Knowledge from Television News,

Newspapers, and News Websites." *International Journal of Communication* 13, (2019): 22.

boyd, d. "White Flight in Networked Publics? How Race and Class Shaped American Teen Engagement with MySpace and Facebook." In *Race After the Internet*, edited by L. Nakamura and P. Chow-White, 203–222. New York: Routledge, 2011.

boyd, d., and J. Heer. "Profiles as Conversation: Networked Identity Performance on Friendster." In *Proceedings of the Hawai'i Int. Conf. on System Sciences (HICSS-39)*, Kauai, HI: IEEE Computer Society, 2006.

boyd, danah, and Eszter Hargittai. "Facebook Privacy Settings: Who Cares?" *First Monday* 15, no. 8 (2010). http://webuse.org/p/a32/.

boyd, danah m., and Nicole B. Ellison. "Social Network Sites: Definition, History, and Scholarship." *Journal of Computer-Mediated Communication* 13, no. 1 (2007): 210–230. https://doi.org/10.1111/j.1083-6101.2007.00393.x.

Brewer, Paul R., and Xiaoxia Cao. "Candidate Appearances on Soft News Shows and Public Knowledge About Primary Campaigns." *Journal of Broadcasting & Electronic Media* 50, no. 1 (2006): 18–35. https://doi.org/10.1207/s15506878jobem5001_2.

Brown, Zoe, and Marika Tiggemann. "Attractive Celebrity and Peer Images on Instagram: Effect on Women's Mood and Body Image." *Body Image* 19 (December 2016): 37–43. https://doi.org/10.1016/j.bodyim.2016.08.007.

Brzezinski, Adam, Guido Deiana, Valentin Kecht, and David Van Dijcke. "The Covid-19 Pandemic: Government vs. Community Action across the United States." *Covid Economics: Vetted and Real-Time Papers* 7 (April 20, 2020): 115–156.

Büchi, Moritz, Noemi Festic, Natascha Just, and Michael Latzer. "Digital Inequalities in Online Privacy Protection: Effects of Age, Education and Gender." In *Handbook of Digital Inequality*, edited by Eszter Hargittai, 296–310. Cheltenham, UK: Edward Elgar, 2021.

Burnes, David, Charles R. Henderson, Christine Sheppard, Rebecca Zhao, Karl Pillemer, and Mark S. Lachs. "Prevalence of Financial Fraud and Scams Among Older Adults in the United States: A Systematic Review and Meta-Analysis." *American Journal of Public Health* 107, no. 8 (2017): e13–e21. https://doi.org/10.2105/AJPH.2017.303821.

Button, Mark, Carol McNaughton Nicholls, Jane Kerr, and Rachael Owen. "Online Frauds: Learning from Victims Why They Fall for These Scams." *Australian & New Zealand Journal of Criminology* 47, no. 3 (2014): 391–408. https://doi.org/10.1177/0004865814521224.

Cacciatore, Michael A., Dietram A. Scheufele, and Elizabeth A. Corley. "Another (Methodological) Look at Knowledge Gaps and the Internet's Potential for Closing Them." *Public Understanding of Science* 23, no. 4 (May 2014): 376–394. https://doi.org/10.1177/0963662512447606.

CBSLA Staff. "LA County Libraries Offer Free Wi-Fi in Parking Lots during COVID-19." CBS Los Angeles, December 23, 2020.

Center for the Study of Hate & Extremism. "Report to the Nation: Anti-Asian Prejudice & Hate Crime." San Bernadino, CA: California State University–San Bernadino, 2021. https://www.csusb.edu/sites/default/files/Report%20to%20the%20 Nation%20-%20Anti-Asian%20Hate%202020%20Final%20Draft%20-%20As%20 of%20Apr%2030%202021%206%20PM%20corrected.pdf.

Chen, W., and B. Wellman. "Charting Digital Divides: Comparing Socioeconomic, Gender, Life Stage, and Rural-Urban Internet Access and Use in Five Countries." In *Transforming Enterprise*, edited by William Dutton, Brian Kahin, Ramon O'Callaghan, and Andrew Wyckoff, 467–497. Cambridge, MA: MIT Press, 2004.

Chipidza, Wallace, Elmira Akbaripourdibazar, Tendai Gwanzura, and Nicole M. Gatto. "Topic Analysis of Traditional and Social Media News Coverage of the Early COVID-19 Pandemic and Implications for Public Health Communication." *Disaster Medicine and Public Health Preparedness*, March 3, 2021, 1–8. https://doi.org /10.1017/dmp.2021.65.

Choi, Wonchan, and Besiki Stvilia. "Web Credibility Assessment: Conceptualization, Operationalization, Variability, and Models." *Journal of the Association for Information Science and Technology* 66, no. 12 (2015): 2399–2414. https://doi.org/10 .1002/asi.23543.

Correa, Teresa, Isabel Pavez, and Javier Contreras. "Digital Inclusion through Mobile Phones?: A Comparison between Mobile-Only and Computer Users in Internet Access, Skills and Use." *Information, Communication & Society* 23, no. 7 (2020): 1074–1091. https://doi.org/10.1080/1369118X.2018.1555270.

Correll, Shelley J. "Gender and the Career Choice Process: The Role of Biased Self-Assessments." *American Journal of Sociology* 106, no. 6 (2001): 1691–1730.

Daube, Elizabeth. "Backlash: Anti-Asian Racism Escalates during COVID-19." *Backlash: Anti–Asian Racism Escalates during COVID-19 | UC San Francisco Magazine*, Summer 2020. https://www.ucsf.edu/magazine/covid-antiasian.

Deurenberg-Yap, M., L. L. Foo, Y. Y. Low, S. P. Chan, K. Vijaya, and M. Lee. "The Singaporean Response to the SARS Outbreak: Knowledge Sufficiency versus Public Trust." *Health Promotion International* 20, no. 4 (2005): 320–326. https://doi .org/10.1093/heapro/dai010.

DeWalt, Darren A., and Michael P. Pignone. "Reading Is Fundamental: The Relationship Between Literacy and Health." *Archives of Internal Medicine* 165, no. 17 (2005): 1943–1944. https://doi.org/10.1001/archinte.165.17.1943.

DiMaggio, Paul, and Eszter Hargittai. "From the 'Digital Divide' to 'Digital Inequality': Studying Internet Use as Penetration Increases." Working Papers. Princeton University, Woodrow Wilson School of Public and International Affairs,

Center for Arts and Cultural Policy Studies, July 2001. https://culturalpolicy
.princeton.edu/sites/culturalpolicy/files/wp15_dimaggio_hargittai.pdf.

Dobransky, Kerry, and Eszter Hargittai. "The Closing Skills Gap: Revisiting the Digital Disability Divide." In *Handbook of Digital Inequality*, edited by Eszter Hargittai, 274–282. Cheltenham, UK: Edward Elgar, 2021.

Dobransky, Kerry, and Eszter Hargittai. "The Disability Divide in Internet Access and Use." *Information, Communication and Society* 9, no. 3 (2006): 313–334. https:// doi.org/10.1080/13691180600751298.

Dobransky, Kerry, and Eszter Hargittai. "People with Disabilities during COVID-19." *Contexts* 19, no. 4 (2020): 46–49. https://doi.org/10.1177/1536504220977935.

Dobransky, Kerry, and Eszter Hargittai. "Piercing the Pandemic Social Bubble: Disability and Social Media Use about COVID-19." *American Behavioral Scientist* 65, no. 12 (2021): 1698–1720. https://doi.org/10.1177/00027642211003146.

Dobransky, Kerry, and Eszter Hargittai. "Unrealized Potential: Exploring the Digital Disability Divide." *Poetics* 58 (2016): 18–28.

Doukas, Charalampos, Vangelis Metsis, Eric Becker, Zhengyi Le, Fillia Makedon, and Ilias Maglogiannis. "Digital Cities of the Future: Extending @home Assistive Technologies for the Elderly and the Disabled | Elsevier Enhanced Reader." *Telematics and Informatics* 28, no. 3 (2011): 176–190. https://doi.org/doi:10.1016/j .tele.2010.08.001.

Duplaga, Mariusz. "Digital Divide among People with Disabilities: Analysis of Data from a Nationwide Study for Determinants of Internet Use and Activities Performed Online." *PLOS ONE* 12, no. 6 (2017): e0179825. https://doi.org/10 .1371/journal.pone.0179825.

Dutton, William H., and Grant Blank. "The Emergence of Next Generation Internet Users." *International Economics and Economic Policy* 11, no. 1–2 (2014): 29–47. https://doi.org/10.1007/s10368-013-0245-8.

Dutton, William H., and Laleah Fernandez. "How Susceptible Are Internet Users?" *Intermedia* 46, no. 4 (December/January, 2019). https://doi.org/10.2139/ssrn.3316768.

Ebrahimji, Alisha. "School Sends California Family a Hotspot after Students Went to Taco Bell to Use Their Free WiFi." *CNN*, September 1, 2020. https://www .cnn.com/2020/08/31/us/taco-bell-california-students-wifi-trnd/index.html.

Elks, Sonia. "Chores and Childcare: Who Bears the Brunt in Lockdown? Women." *Reuters*, May 21, 2020. https://www.reuters.com/article/us-health-coronavirus-women -trfn-idUSKBN22X0G6.

Ellcessor, Elizabeth. *Restricted Access: Media, Disability, and the Politics of Participation*. New York: New York University Press, 2016.

European Commission. "Digital Economy and Society Index (DESI) 2020." 2020. https://digital-strategy.ec.europa.eu/en/policies/desi.

European Commission. "European Accessibility Act." 2019. https://ec.europa.eu /social/main.jsp?catId=1202.

Evans, John H., and Eszter Hargittai. "Who Doesn't Trust Fauci? The Public's Belief in the Expertise and Shared Values of Scientists in the COVID-19 Pandemic." *Socius* 6 (August 6, 2020): 2378023120947337. https://doi.org/10.1177 /2378023120947337.

Evans, John H., and Eszter Hargittai. "Why Would Anyone Distrust Anthony Fauci?" *Scientific American*, June 7, 2020. https://blogs.scientificamerican.com/observations /why-would-anyone-distrust-anthony-fauci/.

Eveland, W.P., and D.A. Scheufele. "Connecting News Media Use with Gaps in Knowledge and Participation." *Political Communication* 17, no. 3 (2000): 215–237. https://doi.org/10.1080/105846000414250.

Evon, Dan. "Was a 'Sacrifice the Weak' Sign Shown at a COVID-19 'ReOpen Tennessee' Rally?" *Snopes*, April 24, 2020. https://www.snopes.com/fact-check /sacrifice-the-weak-sign-real/.

Ferraglioni, Giada. "Dopo un anno di Dad 250 mila studenti senza pc e scuole senza connessione: «Le famiglie si auto-organizzano»." *Open*, March 11, 2021. https:// www.open.online/2021/03/11/covid-19-italia-dad-studenti-senza-pc-e-scuole-senza -connessione/.

Fink, Carsten, and Charles J. Kenny. "W(h)Ither the Digital Divide?" *Info* 5, no. 6 (January 1, 2003): 15–24. https://doi.org/10.1108/14636690310507180.

Fishman, Rachel, and Tamara Hiler. "New Polling from New America & Third Way on COVID-19's Impact on Current and Future College Students." *Third Way*, February 9, 2020.

Fletcher, Richard, and Rasmus Kleis Nielsen. "Are People Incidentally Exposed to News on Social Media? A Comparative Analysis." *New Media & Society* 20, no. 7 (2018): 2450–2468. https://doi.org/10.1177/1461444817724170.

Gamble, V. N. "Under the Shadow of Tuskegee: African Americans and Health Care." *American Journal of Public Health* 87, no. 11 (November 1997): 1773–1778.

Gangadharan, Seeta Peña. "The Downside of Digital Inclusion: Expectations and Experiences of Privacy and Surveillance among Marginal Internet Users." *New Media & Society* 19, no. 4 (2017): 597–615. https://doi.org/10.1177/1461444815614053.

Gaziano, Cecilie. "Knowledge Gap: History and Development." In *The International Encyclopedia of Media Effects*, 1–12. Hoboken, NJ: John Wiley & Sons, 2016. https://doi.org/10.1002/9781118783764.wbieme0041.

Gazit, Tali, Noa Aharony, and Yair Amichai-Hamburger. "Tell Me Who You Are and I Will Tell You Which SNS You Use: SNSs Participation." *Online Information Review* 44, no. 1 (2019): 139–161. https://doi.org/10.1108/OIR-03-2019-0076.

Geiger, A.W. "Key Findings about the Online News Landscape in America." Washington, DC: Pew Research Center, September 11, 2019. https://www.pew research.org/fact-tank/2019/09/11/key-findings-about-the-online-news-landscape -in-america/.

Gerosa, Tiziano, Marco Gui, Eszter Hargittai, and Minh Hao Nguyen. "(Mis)informed during COVID-19: How Education Level and Information Sources Contribute to Knowledge Gaps." *International Journal of Communication* 15 (2021): 22.

Gerson, Jennifer, Anke C. Plagnol, and Philip J. Corr. "Subjective Well-Being and Social Media Use: Do Personality Traits Moderate the Impact of Social Comparison on Facebook?" *Computers in Human Behavior* 63 (2016): 813–822. https://doi .org/10.1016/j.chb.2016.06.023.

Gilardi, Fabrizio. *Digital Technology, Politics, and Policy-Making*. https://www .fabriziogilardi.org/resources/papers/Digital-Technology-Politics-Policy-Making .pdf.

Global Strategy Group. "New America Higher Ed Survey." June 8, 2020. http://thirdway .imgix.net/New-America-and-Third-Way-Higher-Ed-Student-Polling-Data.pdf.

Goffman, Ervin. *The Presentation of Self in Everyday Life*. Garden City, NY: Doubleday, 1959.

Goggin, Gerard. "Disability, Internet, and Digital Inequality: The Research Agenda." In *Handbook of Digital Inequality*, edited by Eszter Hargittai, 255–273. Cheltenham, UK: Edward Elgar, 2021.

Goggin, Gerard M. "Disability and Mobile Internet." *First Monday* 20, no. 9 (September 10, 2015). https://doi.org/10.5210/fm.v20i9.6171.

Goldberg, Rafi. "Digital Divide Among School-Age Children Narrows, but Millions Still Lack Internet Connections." National Telecommunications and Information Administration, December 11, 2018. https://www.ntia.doc.gov/print/blog /2018/digital-divide-among-school-age-children-narrows-millions-still-lack-internet -connections.

Gonzales, Amy. "The Contemporary US Digital Divide: From Initial Access to Technology Maintenance." *Information, Communication & Society* 19, no. 2 (2016): 234–248. https://doi.org/10.1080/1369118X.2015.1050438.

Gonzales, Amy L., Jessica McCrory Calarco, and Teresa Lynch. "Technology Problems and Student Achievement Gaps: A Validation and Extension of the Technology Maintenance Construct." *Communication Research* 47, no. 5 (2020): 750–770. https://doi.org/10.1177/0093650218796366.

Gram, Maggie. "For Many Caregivers and People with Disabilities, WFH Was Never Just a Perk." *New York Times*, May 27, 2020, sec. At Home. https://www .nytimes.com/2020/05/27/at-home/work-from-home-history.html.

Gregory, Sean. "Cory Booker: The Mayor of Twitter and Blizzard Superhero." *Time*, December 29, 2010. http://content.time.com/time/nation/article/0,8599,2039945,00 .html.

Guess, Andrew M., and Benjamin A. Lyons. "Misinformation, Disinformation, and Online Propaganda." In *Social Media and Democracy: The State of the Field, Prospects for Reform*, edited by Nathaniel Persily and Joshua A. Tucker. Cambridge University Press, 2020.

Guess, Andrew, Jonathan Nagler, and Joshua Tucker. "Less than You Think: Prevalence and Predictors of Fake News Dissemination on Facebook." *Science Advances* 5, no. 1 (2019): eaau4586. https://doi.org/10.1126/sciadv.aau4586.

Guess, Andy, Kevin Aslett, Joshua Tucker, Richard Bonneau, and Jonathan Nagler. "Cracking Open the News Feed: Exploring What U.S. Facebook Users See and Share with Large-Scale Platform Data." *Journal of Quantitative Description: Digital Media* 1 (2021). https://doi.org/10.51685/jqd.2021.006.

Guess, Andy, Kevin Munger, Jonathan Nagler, and Joshua Tucker. "How Accurate Are Survey Responses on Social Media and Politics?" *Political Communication* 36, no. 2 (2019): 241–258. https://doi.org/10.1080/10584609.2018.1504840.

Gui, Marco, and Moritz Büchi. "From Use to Overuse: Digital Inequality in the Age of Communication Abundance." *Social Science Computer Review* 39, no. 1 (2021): 3–19. https://doi.org/10.1177/0894439319851163.

Häberli, Stefan. "Coronavirus: Home-Office-Flut bedroht die Schweizer Datennetze." *Neue Zürcher Zeitung*, March 14, 2020. https://www.nzz.ch/wirtschaft /coronavirus-home-office-flut-bedroht-die-schweizer-datennetze-ld.1545722.

Hannon, Taylor. "School Bus Wi-Fi Hotspots Aide Student Learning During COVID-19 Closures." *School Transportation News* (blog), April 8, 2020. https:// stnonline.com/special-reports/school-bus-wi-fi-hotspots-aide-student-learning-dur ing-covid-19-closures/.

Hargittai, Eszter. "Broadband Subsidies Important But More Data Needed to Inform FCC Policy Decisions." *The Huffington Post* (blog), March 28, 2016. http://www .huffingtonpost.com/eszter-hargittai/broadband-subsidies-impor_b_9550960.html.

Hargittai, Eszter. "The Digital Reproduction of Inequality." In *Social Stratification*, edited by David Grusky, 936–944. Boulder, CO: Westview Press, 2008.

Hargittai, Eszter, ed. *Handbook of Digital Inequality*. Cheltenham, UK: Edward Elgar, 2021.

Hargittai, Eszter. "Is Bigger Always Better? Potential Biases of Big Data Derived from Social Network Sites." *The ANNALS of the American Academy of Political and Social Science* 659, no. 1 (2015): 63–76. https://doi.org/10.1177/0002716215570866.

Hargittai, Eszter. "The Making of a Popular Photo App." *Crooked Timber*, March 4, 2016. https://crookedtimber.org/2016/03/04/the-making-of-a-popular-photo-app/.

Hargittai, Eszter. "Potential Biases in Big Data: Omitted Voices on Social Media." *Social Science Computer Review* 38, no. 1 (2020): 10–24. https://doi.org/10.1177/0894439318788322.

Hargittai, Eszter. "Second-Level Digital Divide: Differences in People's Online Skills." *First Monday* 7, no. 4 (April 1, 2002). http://firstmonday.org/ojs/index.php/fm/article/view/942.

Hargittai, Eszter. "Survey Measures of Web-Oriented Digital Literacy." *Social Science Computer Review* 23, no. 3 (2005): 371–379. https://doi.org/10.1177/0894439305275911.

Hargittai, Eszter. "Whose Space? Differences among Users and Non-Users of Social Network Sites." *Journal of Computer-Mediated Communication* 13, no. 1 (2007): 276–297. https://doi.org/10.1111/j.1083-6101.2007.00396.x.

Hargittai, Eszter, and Kerry Dobransky. "Old Dogs, New Clicks: Digital Inequality in Skills and Uses among Older Adults." *Canadian Journal of Communication* 42, no. 2 (2017): 195–212. https://doi.org/10.22230/cjc2017v42n2a3176.

Hargittai, Eszter, and Amanda Hinnant. "Digital Inequality: Differences in Young Adults' Use of the Internet." *Communication Research* 35, no. 5 (2008): 602–621. https://doi.org/10.1177/0093650208321782.

Hargittai, Eszter, and Yuli Patrick Hsieh. "Succinct Survey Measures of Web-Use Skills." *Social Science Computer Review* 30, no. 1 (2012): 95–107. https://doi.org/10.1177/0894439310397146.

Hargittai, Eszter, and Eden Litt. "The Tweet Smell of Celebrity Success: Explaining Variation in Twitter Adoption among a Diverse Group of Young Adults." *New Media & Society* 13, no. 5 (2011): 824–842. https://doi.org/10.1177/1461444811405805.

Hargittai, Eszter, and Marina Micheli. "Internet Skills and Why They Matter." In *Society and the Internet. How Networks of Information and Communication Are Changing Our Lives*, by Mark Graham and William H. Dutton, 109–126, Second Ed. Oxford: Oxford University Press, 2019.

Hargittai, Eszter and Minh Hao Nguyen. "How Switzerland Kept in Touch during Covid-19." Swissinfo.ch, June 19, 2020. https://www.swissinfo.ch/eng/how-people-communicated-during-covid-19-in-switzerland/45848330.

Hargittai, Eszter, Minh Hao Nguyen, Jaelle Fuchs, Jonathan Gruber, Will Marler, Amanda Hunsaker, and Gökçe Karaoglu. "From Zero to a National Data Set in Two Weeks: Reflections on a COVID-19 Collaborative Survey Project." *Social Media + Society* 6, no. 3 (2020): 1–4.

Hargittai, Eszter, Anne Marie Piper, and Meredith Ringel Morris. "From Internet Access to Internet Skills: Digital Inequality among Older Adults." *Universal Access in the Information Society* 18, no. 4 (2018): 881–890. https://doi.org/10.1007/s10209-018-0617-5.

Hargittai, Eszter, and Elissa Redmiles. "Will Americans Be Willing to Install COVID-19 Tracking Apps?" *Scientific American*, April 28, 2020. https://blogs .scientificamerican.com/observations/will-americans-be-willing-to-install-covid-19 -tracking-apps/.

Hargittai, Eszter, Elissa Redmiles, Jessica Vitak, and Michael Zimmer. "Americans' Willingness to Adopt a COVID-19 Tracking App: The Role of App Distributor." *First Monday* 25, no. 11 (2020).

Hargittai, Eszter, and Steven Shafer. "Differences in Actual and Perceived Online Skills: The Role of Gender." *Social Science Quarterly* 87, no. 2 (2006): 432–448. https://doi.org/10.1111/j.1540-6237.2006.00389.x.

Hargittai, Eszter, and Aaron Shaw. "Comparing Internet Experiences and Prosociality in Amazon Mechanical Turk and Population-Based Survey Samples." *Socius* 6 (January 1, 2020): 1–11. https://doi.org/10.1177/2378023119889834.

Hargittai, Eszter, and Florent Thouvenin. "Tracking-App: Die Chancen stehen gut." *Neue Zürcher Zeitung*, May 2, 2020. https://www.nzz.ch/schweiz/tracking -app-chancen-stehen-gut-ld.1554352?reduced=true.

Hargittai, Eszter, and G. Walejko. "The Participation Divide: Content Creation and Sharing in the Digital Age." *Information, Communication and Society* 11, no. 2 (2008): 239–256. https://doi.org/10.1080/13691180801946150.

Hassani, Sara Nephew. "Locating Digital Divides at Home, Work, and Everywhere Else." *Poetics* 34, no. 4–5 (2006): 250–272. https://doi.org/10.1016/j.poetic .2006.05.007.

Hayes, Danny, and Jennifer L. Lawless. "The Decline of Local News and Its Effects: New Evidence from Longitudinal Data." *Journal of Politics* 80, no. 1 (2018): 332–336. https://doi.org/10.1086/694105.

Hellemans, Johan, Kim Willems, and Malaika Brengman. "Daily Active Users of Social Network Sites: Facebook, Twitter, and Instagram-Use Compared to General Social Network Site Use." In *Advances in Digital Marketing and ECommerce*, edited by Francisco J. Martínez-López and Steven D'Alessandro, 194–202. Springer Proceedings in Business and Economics. Cham, Switzerland: Springer International, 2020. https://doi.org/10.1007/978-3-030-47595-6_24.

Hermida, Alfred, Fred Fletcher, Darryl Korell, and Donna Logan. "Share, Like, Recommend: Decoding the Social Media News Consumer." *Journalism Studies* 13, no. 5–6 (October 2012): 815–824. https://doi.org/10.1080/1461670X.2012.664430.

Ho, Shirley S. "The Knowledge Gap Hypothesis in Singapore: The Roles of Socioeconomic Status, Mass Media, and Interpersonal Discussion on Public Knowledge of the H1N1 Flu Pandemic." *Mass Communication and Society* 15, no. 5 (2012): 695–717. https://doi.org/10.1080/15205436.2011.616275.

Hofer, Matthias, Eszter Hargittai, Moritz Büchi, and Alexander Seifert. "Older Adults' Online Information Seeking and Subjective Well-Being: The Moderating

Role of Internet Skills." *International Journal of Communication* 13 (September 13, 2019): 4426–4443.

Hoffman, Donna L., Thomas P. Novak, and Ann Schlosser. "The Evolution of the Digital Divide: How Gaps in Internet Access May Impact Electronic Commerce." *Journal of Computer-Mediated Communication* 5, no. 3 (2000): JCMC534. https://doi .org/10.1111/j.1083-6101.2000.tb00341.x.

Hollier, Scott. "The Growing Importance of Accessible Social Media." In *Disability and Social Media: Global Perspectives*, edited by Katie Ellis and Mike Kent, 77–88. Interdisciplinary Disability Studies. New York: Routledge, 2017.

Holton, Bill. "Facebook Accessibility for Users with Visual Impairments: What Facebook Wants You to Know." *The American Foundation for the Blind*, April 2015. https://www.afb.org/aw/16/4/15469.

Hoskin, Janet, and Jo Finch. "How Disabled People Have Been Completely Disregarded during the Coronavirus Pandemic." *The Conversation*, July 27, 2020. https:// theconversation.com/how-disabled-people-have-been-completely-disregarded-dur ing-the-coronavirus-pandemic-142766.

Hsieh, Yuli Patrick. "Online Social Networking Skills: The Social Affordances Approach to Digital Inequality." *First Monday* 17, no. 4 (2012). https://doi.org/10 .5210/fm.v17i4.3893.

Huber, Stephan Gerhard, Paula Sophie Günther, Nadine Schneider, Christoph Helm, Marius Schwander, Julia Schneider, and Jane Pruitt. *COVID-19 und aktuelle Herausforderungen in Schule und Bildung*. Münster, Germany: Waxmann Verlag GmbH, 2020. https://doi.org/10.31244/9783830942160.

Hunsaker, Amanda, and Eszter Hargittai. "Age-Related Differences in Home Experiences and Worries during COVID-19." Working paper. December 5, 2020.

Hunsaker, Amanda, Minh Hao Nguyen, Jaelle Fuchs, Gökçe Karaoglu, Teodora Djukaric, and Eszter Hargittai. "Unsung Helpers: Older Adults as a Source of Digital Media Support for Their Peers." *The Communication Review* 23, no. 4 (2020): 309–330. https://doi.org/10.1080/10714421.2020.1829307.

Hwang, Yoori, and Se-Hoon Jeong. "Revisiting the Knowledge Gap Hypothesis: A Meta-Analysis of Thirty-Five Years of Research—Yoori Hwang, Se-Hoon Jeong, 2009." *Journalism & Mass Communication Quarterly* 86, no. 3 (2009): 513–532.

Igielnik, Ruth. "A Rising Share of Working Parents in the U.S. Say It's Been Difficult to Handle Child Care during the Pandemic." Washington, DC: Pew Research Center, 2021. https://www.pewresearch.org/fact-tank/2021/01/26/a-rising-share-of -working-parents-in-the-u-s-say-its-been-difficult-to-handle-child-care-during-the -pandemic/.

il Fatto Nisseno Editorial Board. "Coronavirus, Orlando to Conte: 'Guarantee All Citizens. Unlimited and Free Wi-Fi for the Duration of the Emergency.'" *il Fatto Nisseno*, April 3, 2020. https://www.ilfattonisseno.it/2020/04/coronavirus

-orlando-a-conte-garantire-tutti-i-cittadini-wi-fi-illimitato-e-gratuito-per-tutta
-la-durata-dellemergenza/.

Ingen, Eric van, and Uwe Matzat, "Inequality in Mobilizing Online Help after a Negative Life Event: The Role of Education, Digital Skills, and Capital-Enhancing Internet Use." *Information, Communication & Society* 21, no. 4 (2018): 481–498. https://doi.org/10.1080/1369118X.2017.1293708.

Italian National Institute of Statistics. "Multiscopo ISTAT—Aspects of Daily Life (2016)." Unidata, 2018. https://tinyurl.com/y4kl554e.

Jackson, Sarah J., Moya Bailey, and Brooke Foucault Welles. *#HashtagActivism: Networks of Race and Gender Justice.* Cambridge, MA: MIT Press, 2020. https://mitpress.mit.edu/books/hashtagactivism.

Jaiswal, J., C. LoSchiavo, and D. C. Perlman. "Disinformation, Misinformation and Inequality-Driven Mistrust in the Time of COVID-19: Lessons Unlearned from AIDS Denialism." *AIDS and Behavior* 24, no. 10 (2020): 2776–2780. https://doi.org/10.1007/s10461-020-02925-y.

Jeffres, Leo, Kimberly Neuendorf, and David Atkin. "Acquiring Knowledge From the Media in the Internet Age." *Communication Quarterly* 60 (January 1, 2012): 59–79. https://doi.org/10.1080/01463373.2012.641835.

Jenssen, Anders Todal. "Widening or Closing the Knowledge Gap?: The Role of TV and Newspapers in Changing the Distribution of Political Knowledge." *Nordicom Review* 33, no. 1 (2013): 19–36. https://doi.org/10.2478/nor-2013-0002.

Jurkowitz, M., and A. Mitchell. "Americans Who Get News Mostly through Social Media Are Least Likely to Follow Coronavirus Coverage." *Pew Research Center's Journalism Project* (blog), March 25, 2020. https://www.journalism.org/2020/03/25/americans-who-primarily-get-news-through-social-media-are-least-likely-to-follow-covid-19-coverage-most-likely-to-report-seeing-made-up-news/.

Kang, Cecilia. "Parking Lots Have Become a Digital Lifeline." *The New York Times*, May 5, 2020, sec. Technology.

Kappeler, Kiran, Noemi Festic, and Michael Latzer. "Who Remains Offline and Why? Growing Social Stratification of Internet Use in the Highly Digitized Swiss Society." Zurich: University of Zurich, 2020. https://mediachange.ch/media//pdf/publications/nonuse.pdf.

Karpf, David. "Social Science Research Methods in Internet Time." *Information, Communication & Society* 15, no. 5 (2012): 639–661. https://doi.org/10.1080/1369118X.2012.665468.

Katz, J. E., and R. E. Rice. *Social Consequences of Internet Use: Access, Involvement and Interaction.* Cambridge, MA: MIT Press, 2002. https://mitpress.mit.edu/books/social-consequences-internet-use.

Katz, James E., Ronald E. Rice, and Philip Aspden. "The Internet, 1995–2000: Access, Civic Involvement, and Social Interaction." *American Behavioral Scientist* 45, no. 3 (2001): 405–419. https://doi.org/10.1177/0002764201045003004.

Kelén, Joana, Ruth Fulterer, and Barnaby Skinner. "Coronavirus: So Infizierten Sich Die Ersten 50 Schweizer." *NZZ*, June 3, 2020. https://www.nzz.ch/schweiz /coronavirus-so-infizierten-sich-die-ersten-50-schweizer-ld.1544556.

Kent, Rachael. "Self-Tracking Health Over Time: From the Use of Instagram to Perform Optimal Health to the Protective Shield of the Digital Detox." *Social Media + Society* 6, no. 3 (2020): 2056305120940694. https://doi.org/10.1177/2056 305120940694.

Kessler, Sabrina. "Coronavirus deckt Mängel auf—«Mach deine Hausaufgaben vom Parkplatz aus»." Schweizer Radio und Fernsehen (SRF), June 1, 2020. https://www.srf.ch/news/international/coronavirus-deckt-maengel-auf-mach -deine-hausaufgaben-vom-parkplatz-aus.

Klopfenstein, Bruce C. "Audience Measurement in the VCR Environment: An Examinatino of Ratings Methodologies." In *Social & Cultural Aspects of VCR Use*, edited by Julia R. Dobrow, 45–69. Hillsdale, NJ: Lawrence Erlbaum Associates, Inc, 1990. https://books.google.ch/books?id=kK934crW140C&lpg=PA45&ots =hLZRTiMSOx&dq=review%20peoplemeter%20data&lr&pg=PA45#v=onepage &q&f=false.

König, Ronny, Alexander Seifert, and Michael Doh. "Internet Use among Older Europeans: An Analysis Based on SHARE Data." *Universal Access in the Information Society* 17, no. 3 (2018): 621–633. https://doi.org/10.1007/s10209-018-0609-5.

Koumpilova, Mila. "Chicago Helped 55,000 Students Get Free Internet. Much Work Remains." *Chalkbeat Chicago*, January 12, 2021. https://chicago.chalkbeat .org/2021/1/12/22227642/chicago-helped-55000-students-get-free-internet-much -work-remains.

Koumpilova, Mila. "Chicago Is Spending $50M to Get Low-Income Students Online. What If Parents Don't Trust a Free Deal?" *Chalkbeat Chicago*, September 9, 2020. https://chicago.chalkbeat.org/2020/9/9/21428427/another-hurdle-for -chicagos-internet-push-reluctance-to-take-free-deal.

Lancioni, Giulio E., Jeff Sigafoos, Mark F. O'Reilly, and Nirbhay N. Singh. "Speech-Generating Devices for Communication and Social Development." In *Assistive Technology: Interventions for Individuals with Severe/Profound and Multiple Disabilities*, edited by Giulio E. Lancioni, Jeff Sigafoos, Mark F. O'Reilly, and Nirbhay N. Singh, 41–71. Autism and Child Psychopathology Series. New York: Springer, 2013. https://doi.org/10.1007/978-1-4614-4229-5_3.

Lawler, Dave. "Timeline: How Italy's Coronavirus Crisis Became the World's Deadliest." *Axios*, March 24, 2020. https://www.axios.com/italy-coronavirus-timeline -lockdown-deaths-cases-2adb0fc7-6ab5-4b7c-9a55-bc6897494dc6.html.

Lazar, Jonathan, and Paul T. Jaeger. "Reducing Barriers to Online Access for People with Disabilities." *Issues in Science and Technology* 27, no. 2 (2011): 612–630.

Lazer, David, Eszter Hargittai, Deen Freelon, Sandra Gonzalez-Bailon, Kevin Munger, Katherine Ognyanova, and Jason Radford. "Meaningful Measures of Human Society in the Twenty-First Century." *Nature* 595 (2021): 189–196. https://doi.org/10.1038/s41586-021-03660-7.

Light, Ben, Jean Burgess, and Stefanie Duguay. "The Walkthrough Method: An Approach to the Study of Apps." *New Media & Society*, 20, no. 3 (2016): 881–900. https://doi.org/10.1177/1461444816675438.

Lind, Fabienne, and Hajo G. Boomgaarden. "What We Do and Don't Know: A Meta-Analysis of the Knowledge Gap Hypothesis." *Annals of the International Communication Association* 43, no. 3 (2019): 210–224. https://doi.org/10.1080/23808985.2019.1614475.

Litt, Eden, and Eszter Hargittai. "'Just Cast the Net, and Hopefully the Right Fish Swim into It': Audience Management on Social Network Sites." In *Proceedings of the 19th ACM Conf. on Computer-Supported Cooperative Work & Social Computing*, 1488–1500. San Francisco: ACM, 2016. https://doi.org/10.1145/2818048.2819933.

Litt, Eden, Erin Spottswood, Jeremy Birnholtz, Jeff T. Hancock, Madeline E. Smith, and Lindsay Reynolds. "Awkward Encounters of an 'Other' Kind: Collective Self-Presentation and Face Threat on Facebook." *Proceedings of the 17th ACM Conf. on Computer-Supported Cooperative Work & Social Computing*, 449–460. New York: ACM, 2014. https://doi.org/10.1145/2531602.2531646.

Livingstone, Sonia, and Ellen Helsper. "Gradations in Digital Inclusion: Children, Young People and the Digital Divide." *New Media & Society* 9, no. 4 (2007): 671–696. https://doi.org/10.1177/1461444807080335.

Marler, Will, Eszter Hargittai, and Minh Hao Nguyen. "Can You See Me Now? Video Gatherings and Social Connectedness during the COVID-19 Pandemic." *The Information Society* 38, no. 1 (2022): 36–50. https://doi.org/10.1080/01972243.2021.2007193.

Marwick, Alice E. "To Catch a Predator? The MySpace Moral Panic." *First Monday* 13, no. 6 (2008). https://doi.org/10.5210/fm.v13i6.2152.

Marwick, Alice E. "Why Do People Share Fake News? A Sociotechnical Model of Media Effects." *Georgetown Law Technology Review 2* (2018): 474–512.

Marwick, Alice E., and danah boyd. "Networked Privacy: How Teenagers Negotiate Context in Social Media." *New Media & Society* 16, no. 7 (2014): 1051–1067. https://doi.org/10.1177/1461444814543995.

Mascheroni, Giovanna, Marium Saeed, Marco Valenza, Davide Cino, and Thomas Dreesen. "La didattica a distanza durante l'emergenza COVID-19: l'esperienza italiana." Rome: UNICEF, 2021.

McGonagle, Tarlach. "'Fake News': False Fears or Real Concerns?" *Netherlands Quarterly of Human Rights* 35, no. 4 (2017): 203–209. https://doi.org/10.1177 /0924051917738685.

McLaughlin, Caitlin, and Jessica Vitak. "Norm Evolution and Violation on Facebook." *New Media & Society* 14, no. 2 (2012): 299–315. https://doi.org/10.1177 /1461444811412712.

McQuail, Denis. *Audience Analysis*. London: SAGE, 1997.

Meiselwitz, Gabriele, Brian Wentz, and Jonathan Lazar. "Universal Usability: Past, Present, and Future." *Foundations and Trends in Human–Computer Interaction* 3, no. 4 (2010): 213–333. https://doi.org/10.1561/1100000029.

Mellon, Jonathan, and Christopher Prosser. "Twitter and Facebook Are Not Representative of the General Population: Political Attitudes and Demographics of British Social Media Users." *Research & Politics*, July 13, 2017. https://doi.org/10 .1177/2053168017720008.

Metzger, Miriam J. "Making Sense of Credibility on the Web: Models for Evaluating Online Information and Recommendations for Future Research." *Journal of the American Society for Information Science and Technology* 58, no. 13 (2007): 2078–2091. https://doi.org/10.1002/asi.20672.

Micheli, Marina. "Social Networking Sites and Low-Income Teenagers: Between Opportunity and Inequality." *Information, Communication & Society* 19, no. 5 (2016): 565–581. https://doi.org/10.1080/1369118X.2016.1139614.

Micheli, Marina, Elissa M. Redmiles, and Eszter Hargittai. "Help Wanted: Young Adults' Sources of Support for Questions about Digital Media." *Information, Communication & Society* 23, no. 11 (2020): 1655–1672. https://doi.org/10.1080/1369118X .2019.1602666.

Michikyan, Minas, Kaveri Subrahmanyam, and Jessica Dennis. "Can You Tell Who I Am? Neuroticism, Extraversion, and Online Self-Presentation among Young Adults." *Computers in Human Behavior* 33 (April 1, 2014): 179–183. https://doi.org /10.1016/j.chb.2014.01.010.

Milne, George R., Lauren I. Labrecque, and Cory Cromer. "Toward an Understanding of the Online Consumer's Risky Behavior and Protection Practices." *Journal of Consumer Affairs* 43, no. 3 (2009): 449–473. https://doi.org/10.1111/j .1745-6606.2009.01148.x.

Ministero dell'Economia e delle Finanze. "I provvedimenti del Governo a sostegno del Lavoro." September 28, 2020. https://www.mef.gov.it/covid-19/I-provvedimenti -del-Governo-a-sostegno-del-Lavoro/.

Mitchell, Amy, Mark Jurkowitz, J. Baxter Oliphant, and Elisa Shearer. "How Americans Navigated the News in 2020: A Tumultuous Year in Review." Washington, DC: Pew Research Center, February 22, 2021. https://www.journalism

.org/2021/02/22/how-americans-navigated-the-news-in-2020-a-tumultuous-year
-in-review/.

Mitchell, Amy, J. Baxter Oliphant, and Elisa Shearer. "About Seven-in-Ten U.S. Adults Say They Need to Take Breaks From COVID-19 News." *Pew Research Center's Journalism Project* (blog), April 29, 2020. https://www.journalism.org/2020/04/29 /about-seven-in-ten-u-s-adults-say-they-need-to-take-breaks-from-covid-19-news/.

Morris, David S., and Jonathan S. Morris. "Evolving Learning: The Changing Effect of Internet Access on Political Knowledge and Engagement (1998–2012)." *Sociological Forum* 32, no. 2 (2017): 339–358. https://doi.org/10.1111/socf.12333.

Morrow, Robert W. *"Sesame Street" and the Reform of Children's Television*. Baltimore: Johns Hopkins University Press, 2006. https://jhupbooks.press.jhu.edu /title/sesame-street-and-reform-childrens-television.

Mossberger, Karen, Caroline J. Tolbert, and Mary Stansbury. *Virtual Inequality: Beyond the Digital Divide*. Washington, DC: Georgetown University Press, 2003.

Motta, Matt, Dominik Stecula, and Christina Farhart. "How Right-Leaning Media Coverage of COVID-19 Facilitated the Spread of Misinformation in the Early Stages of the Pandemic in the U.S." *Canadian Journal of Political Science/Revue Canadienne de Science Politique* 53, no. 2 (June 2020): 335–342. https://doi.org/10 .1017/S0008423920000396.

National Association of Black Journalists. "NABJ Style Guide—National Association of Black Journalists." *NABJ Statement on Capitalizing Black and Other Racial Identifiers* (blog), June 2020. https://www.nabj.org/page/styleguide.

National Telecommunications and Information Administration. "Falling Through the Net: A Survey of the 'Have Nots' in Rural and Urban America." Washington, DC: US Dept. of Commerce, 1995. http://www.ntia.doc.gov/ntiahome /fallingthru.html#2/.

Natriello, Gary. "Bridging the Second Digital Divide: What Can Sociologists of Education Contribute?" *Sociology of Education* 74, no. 3 (2001): 260–265.

Nguyen, Minh Hao, Jonathan Gruber, Jaelle Fuchs, Will Marler, Amanda Hunsaker, and Eszter Hargittai. "Changes in Digital Communication during the COVID-19 Global Pandemic: Implications for Digital Inequality and Future Research." *Social Media + Society* 6, no. 3 (2020). https://doi.org/10.1177/2056305120948255.

Nguyen, Minh Hao, Jonathan Gruber, Will Marler, Amanda Hunsaker, Jaelle Fuchs, and Eszter Hargittai. "Staying Connected while Physically Apart: Digital Communication when Face-to-Face Interactions Are Limited." *New Media & Society* (February 2021). https://doi.org/10.1177/1461444820985442.

Nguyen, Minh Hao, Eszter Hargittai, Jaelle Fuchs, Teodora Djukaric, and Amanda Hunsaker. "Trading Spaces: How and Why Older Adults Disconnect from and Switch between Digital Media." *The Information Society* 37, no. 5 (2021): 1–13. https://doi.org/10.1080/01972243.2021.1960659.

Nguyen, Minh Hao, Eszter Hargittai, and Will Marler. "Digital Inequality in Communication during a Time of Physical Distancing: The Case of COVID-19." *Computers in Human Behavior* 120 (2021). https://doi.org/10.1016/j.chb.2021.106717.

Norris, Pippa. *Digital Divide: Civic Engagement, Information Poverty and the Internet in Democratic Societies.* New York: Cambridge University Press, 2001.

OECD. "Crisis Squeezes Income and Puts Pressure on Inequality and Poverty." 2013.

Ono, Hiroshi, and Takeshi Mori. "COVID-19 and Telework: An International Comparison." *Journal of Quantitative Description: Digital Media* 1 (April 26, 2021). https://doi.org/10.51685/jqd.2021.004.

Ono, Hiroshi, and Madeline Zavodny. "Digital Inequality: A Five Country Comparison Using Microdata." *Social Science Research* 36, no. 3 (2007): 1135–1155. https://doi.org/10.1016/j.ssresearch.2006.09.001.

Palca, Joe. "NIH Panel Recommends Against Drug Combination Promoted by Trump for COVID-19." *NPR*, April 21, 2020, sec. The Coronavirus Crisis. https://www.npr.org/sections/coronavirus-live-updates/2020/04/21/840341224/nih-panel-recommends-against-drug-combination-trump-has-promoted-for-covid-19.

Pew Research Center. "Demographics of Social Media Users and Adoption in the United States." *Pew Research Center: Internet, Science & Tech* (blog), April 7, 2021. https://www.pewresearch.org/internet/fact-sheet/social-media/.

Pew Research Center. "Internet/Broadband Fact Sheet." 2021. https://www.pewresearch.org/internet/fact-sheet/internet-broadband/.

Pew Research Center. "Republicans, Democrats Move Even Further Apart in Coronavirus Concerns." Washington, DC: Pew Research Center, June 25, 2020. https://www.pewresearch.org/politics/2020/06/25/republicans-democrats-move-even-further-apart-in-coronavirus-concerns/.

Prasad, Amit. "Anti-Science Misinformation and Conspiracies: COVID–19, Post-Truth, and Science & Technology Studies (STS)." *Science, Technology and Society* 27, no. 1 (2022): 88–112. https://doi.org/10.1177/09717218211003413.

Prensky, M. "Digital Natives, Digital Immigrants." *On the Horizon* 9, no. 5 (2001). http://www.marcprensky.com/writing/Prensky%20-%20Digital%20Natives,%20Digital%20Immigrants%20-%20Part1.pdf.

Quan-Haase, Anabel, Carly Williams, Maria Kicevski, Isioma Elueze, and Barry Wellman. "Dividing the Grey Divide: Deconstructing Myths About Older Adults' Online Activities, Skills, and Attitudes." *American Behavioral Scientist* 62, no. 9 (2018): 1207–1228. https://doi.org/10.1177/0002764218777572.

Rai Scuola. "La Scuola in tivù." Rai Scuola. Accessed July 15, 2021. https://www.raiscuola.rai.it/raiscuola/articoli/2021/01/La-Scuola-in-TV-gli-orari-delle-lezioni-ecc87dfb-e9d6-43c8-8c45-1d69d2ec8c17.html.

Razzi, Massimo. "La tecnologia è un bene comune, ma più del 12% degli studenti italiani è senza computer." *la Repubblica*, July 20, 2020. https://www.repubblica.it/solidarieta/diritti-umani/2020/07/20/news/disuguaglianze_digitali-262453676/.

Redmiles, Elissa M., and Cody L. J. Buntain. "How Feelings of Trust, Concern, and Control of Personal Online Data Influence Web Use." In *Handbook of Digital Inequality*, edited by Eszter Hargittai, 311–325. Cheltenham, UK: Edward Elgar, 2021.

Redmiles, Elissa M., Gabriel Kaptchuk, and Eszter Hargittai. "The Success of Contact Tracing Doesn't Just Depend on Privacy." *Wired*, May 23, 2020. https://www.wired.com/story/the-success-of-contact-tracing-doesnt-just-depend-on-privacy/.

Redmiles, Elissa M., Sean Kross, and Michelle L. Mazurek. "Where Is the Digital Divide? A Survey of Security, Privacy, and Socioeconomics." *CHI '17: Proceedings of the 2017 CHI Conference on Human Factors in Computing Systems*, 931–936. New York: ACM, May 2017. https://doi.org/10.1145/3025453.3025673.

Refle, Jan-Erik, and Marieke Voorpostel. "First Results of the Swiss Household Panel—Covid-19 Study." 2020. https://doi.org/doi:10.24440/FWP-2020-00001.

Robinson, Laura, Shelia R. Cotten, Hiroshi Ono, Anabel Quan-Haase, Gustavo Mesch, Wenhong Chen, Jeremy Schulz, Timothy M. Hale, and Michael J. Stern. "Digital Inequalities and Why They Matter." *Information, Communication & Society* 18, no. 5 (2015): 569–582. https://doi.org/10.1080/1369118X.2015.1012532.

Rogers, Chris, and Noelle Ellerson Ng. "Report of Initial Findings: COVID-19 Impact on Public Schools." AASA—The School Superintendents Association, March 27, 2020. https://www.aasa.org/uploadedFiles/AASA_Blog(1)/AASA%20COVID%20survey%20INITIAL%20003272020%20FN.pdf.

Roser, Max, Hannah Ritchie, and Esteban Ortiz-Ospina. "Internet." Our World in Data. Oxford: University of Oxford, July 14, 2015. https://ourworldindata.org/internet.

Schlomann, Anna, Alexander Seifert, Susanne Zank, and Christian Rietz. "Assistive Technology and Mobile ICT Usage Among Oldest-Old Cohorts: Comparison of the Oldest-Old in Private Homes and in Long-Term Care Facilities." *Research on Aging* 42, no. 5–6 (2020): 163–173. https://doi.org/10.1177/0164027520911286.

Schradie, Jen. "The Digital Production Gap: The Digital Divide and Web 2.0 Collide." *Poetics* 39, no. 2 (2011): 145–168. https://doi.org/10.1016/j.poetic.2011.02.003.

Schwei, Rebecca J., Kelley Kadunc, Anthony L. Nguyen, and Elizabeth A. Jacobs. "Impact of Sociodemographic Factors and Previous Interactions with the Health Care System on Institutional Trust in Three Racial/Ethnic Groups." *Patient Education and Counseling* 96, no. 3 (September 2014): 333–338. https://doi.org/10.1016/j.pec.2014.06.003.

Selwyn, Neil. "Defining the 'Digital Divide': Developing a Theoretical Understanding of Inequalities in the Information Age." Occasional Paper 49. Cardiff,

Wales: Cardiff University, 2002. http://ictlogy.net/bibliography/reports/projects
.php?idp=348.

Sevilla, Almudena, and Sarah Smith. "Baby Steps: The Gender Division of Childcare
during the COVID-19 Pandemic." *Oxford Review of Economic Policy* 36, no. Supplement 1 (September 28, 2020): S169–S186. https://doi.org/10.1093/oxrep/graa027.

Shapiro, Joseph. "Oregon Hospitals Didn't Have Shortages. So Why Were Disabled People Denied Care?" *NPR.org*, December 21, 2020. https://text.npr.org
/946292119.

Shapiro, Joseph. "People With Disabilities Say Rationing Care Policies Violate
Civil Rights." *NPR.org*, March 23, 2020. https://text.npr.org/820398531.

Sharma, Megha, Kapil Yadav, Nitika Yadav, and Keith C. Ferdinand. "Zika Virus
Pandemic—Analysis of Facebook as a Social Media Health Information Platform."
American Journal of Infection Control 45, no. 3 (2017): 301–302. https://doi.org/10
.1016/j.ajic.2016.08.022.

Shaw, Aaron, and Eszter Hargittai. "The Pipeline of Online Participation Inequalities: The Case of Wikipedia Editing." *Journal of Communication* 68, no. 1 (2018):
143–168. https://doi.org/10.1093/joc/jqx003.

Sheldon, Pavica, and Katherine Bryant. "Instagram: Motives for Its Use and Relationship to Narcissism and Contextual Age." *Computers in Human Behavior* 58, no.
Supplement C (May 1, 2016): 89–97. https://doi.org/10.1016/j.chb.2015.12.059.

Smit, Edith G., Guda Van Noort, and Hilde A. M. Voorveld. "Understanding
Online Behavioural Advertising: User Knowledge, Privacy Concerns and Online
Coping Behaviour in Europe." *Computers in Human Behavior* 32 (March 2014): 15–
22. https://doi.org/10.1016/j.chb.2013.11.008.

Smith, Marc A., and Peter Kollock, eds. *Communities in Cyberspace*. London: Routledge, 1999.

Spence, Patric R., Kenneth A. Lachlan, and Jennifer A. Burke. "Differences in Crisis
Knowledge across Age, Race, and Socioeconomic Status during Hurricane Ike: A
Field Test and Extension of the Knowledge Gap Hypothesis." *Communication Theory*
21, no. 3 (2011): 261–278. https://doi.org/10.1111/j.1468-2885.2011.01385.x.

Steenhuysen, Julie, and Dan Whitcomb. "Washington State Man Who Traveled to
China Is First U.S. Victim of Coronavirus." *Reuters*, January 21, 2020, sec. Healthcare &
Pharma. https://www.reuters.com/article/us-china-health-usa-idUSKBN1ZK2FF.

Stelitano, Laura, Sy Doan, Ashley Woo, Melissa Kay Diliberti, Julia H. Kaufman,
and Daniella Henry. "The Digital Divide and COVID-19: Teachers' Perceptions
of Inequities in Students' Internet Access and Participation in Remote Learning."
RAND Corporation, September 24, 2020. https://www.rand.org/pubs/research
_reports/RRA134-3.html.

Stern, Christopher. "New FCC Chairman Favors a Non-Activist Approach." *Washington Post*, February 7, 2001. https://www.washingtonpost.com/archive/business

/2001/02/07/new-fcc-chairman-favors-a-non-activist-approach/1e1b5725-48b2
-4bda-85ac-31fdca2d1917/.

Sullivan, Laura Specker. "Trust, Risk, and Race in American Medicine." *Hastings Center Report* 50, no. 1 (2020): 18–26. https://doi.org/10.1002/hast.1080.

Sunstein, Cass. "Deciding by Default." *University of Pennsylvania Law Review* 162, no. 1 (December 1, 2013): 1–57.

Swiss Confederation. "Short-Time Working." Accessed July 15, 2021. https://www.ch.ch/en/short-time-work/.

"Television History—The First 75 Years." Accessed July 19, 2021. http://www.tvhistory.tv/facts-stats.htm.

Tichenor, Philipp J., George A. Donohue, and Clarice N. Olien. "Mass Media Flow and Differential Growth in Knowledge." *The Public Opinion Quarterly* 34, no. 2 (1970): 159–170.

Tiggemann, Marika, and Isabella Anderberg. "Sòcial Media Is Not Real: The Effect of 'Instagram vs Reality' Images on Women's Social Comparison and Body Image." *New Media & Society* 22, no. 12 (2020): 2183–2199. https://doi.org/10.1177/1461444819888720.

Tillmann, R., M. Voorpostel, U. Kuhn, F. Lebert, V.-A. Ryser, O. Lipps, B. Wernli, and E. Antal. "The Swiss Household Panel Study: Observing Social Change since 1999." *Longitudinal and Life Course Studies* 7, no. 1 (2016): 64–78. https://doi.org/10.14301/llcs.v7i1.360.

Treem, Jeffrey W., Stephanie L. Dailey, Casey S. Pierce, and Diana Biffl. "What We Are Talking About When We Talk About Social Media: A Framework for Study." *Sociology Compass* 10, no. 9 (2016): 768–784. https://doi.org/10.1111/soc4.12404.

Treem, Jeffrey W., and Paul M. Leonardi. "Social Media Use in Organizations: Exploring the Affordances of Visibility, Editability, Persistence, and Association." *Annals of the International Communication Association* 36, no. 1 (2013): 143–189. https://doi.org/10.1080/23808985.2013.11679130.

Tufekci, Zeynep. "Can You See Me Now? Audience and Disclosure Regulation in Online Social Network Sites." *Bulletin of Science, Technology & Society* 28, no. 20 (2008): 20–36. https://doi.org/10.1177/0270467607311484.

Tufekci, Zeynep. "Grooming, Gossip, Facebook and Myspace." *Information, Communication & Society* 11, no. 4 (2008): 544–564. https://doi.org/10.1080/13691180801999050.

Tufekci, Zeynep. "How Social Media Took Us from Tahrir Square to Donald Trump." *MIT Technology Review*, August 14, 2018. https://www.technologyreview.com/2018/08/14/240325/how-social-media-took-us-from-tahrir-square-to-donald-trump/.

"2010 ADA Regulations." US Dept. of Justice, Civil Rights Division, 2017. https://www.ada.gov/2010_regs.htm.

United States Census Bureau. "Current Population Survey (CPS)." 2021. https://www.census.gov/programs-surveys/cps.html.

US Government Publishing Office. US Rehabilitation Act of 1973, 29 § 508 (1973).

Vagnoni, Giselda. "Researchers Find Coronavirus Was Circulating in Italy Earlier than Thought." *Reuters*, November 16, 2020. https://www.reuters.com/article/health-coronavirus-italy-timing-idUSKBN27W1J2.

Vaidhyanathan. "Generational Myth." *Chronicle of Higher Education* 55, no. 4 (2008): B7–B9.

van Deursen, Alexander J. A. M. "Internet Skill-Related Problems in Accessing Online Health Information." *International Journal of Medical Informatics* 81, no. 1 (2011): 61–72. https://doi.org/10.1016/j.ijmedinf.2011.10.005.

van Deursen, Alexander J.A.M. "Internet Skills: Vital Assets in an Information Society." PhD thesis, University of Twente, 2010. https://doi.org/10.3990/1.9789036530866.

van Deursen, Alexander J. A. M., and Jan A. G. M. van Dijk. "Internet Skills Performance Tests: Are People Ready for EHealth?" *Journal of Medical Internet Research* 13, no. 2 (2011): e35. https://doi.org/10.2196/jmir.1581.

van Deursen, Alexander J. A. M., Jan A. G. M. van Dijk, and Peter M. ten Klooster. "Increasing Inequalities in What We Do Online: A Longitudinal Cross Sectional Analysis of Internet Activities among the Dutch Population (2010 to 2013) over Gender, Age, Education, and Income." *Telematics and Informatics* 32, no. 2 (2015): 259–272. https://doi.org/10.1016/j.tele.2014.09.003.

van Dijck, José. *The Culture of Connectivity: A Critical History of Social Media*. New York: Oxford University Press, 2013.

van Dijk, Jan A. G. M. "Widening Information Gaps and Policies of Prevention." In *Digital Democracy*, edited by Kenneth L. Hacker and Jan A. G. M. van Dijk, 166–183. London: SAGE, 2000.

Vicente, María Rosalía, and Ana Jesús López. "A Multidimensional Analysis of the Disability Digital Divide: Some Evidence for Internet Use." *The Information Society* 26, no. 1 (2010): 48–64. https://doi.org/10.1080/01615440903423245.

Viner, Russell M, Aswathikutty Gireesh, Neza Stiglic, Lee D Hudson, Anne-Lise Goddings, Joseph L Ward, and Dasha E Nicholls. "Roles of Cyberbullying, Sleep, and Physical Activity in Mediating the Effects of Social Media Use on Mental Health and Wellbeing among Young People in England: A Secondary Analysis of Longitudinal Data." *Lancet Child & Adolescent Health* 3, no. 10 (2019): 685–696. https://doi.org/10.1016/S2352-4642(19)30186-5.

Viswanath, K., and John R. Jr. Finnegan. "The Knowledge Gap Hypothesis: Twenty-Five Years Later." *Annals of the International Communication Association* 19, no. 1 (1996): 187–227.

Vitak, Jessica. "The Impact of Context Collapse and Privacy on Social Network Site Disclosures." *Journal of Broadcasting & Electronic Media* 56, no. 4 (2012): 451–470. https://doi.org/10.1080/08838151.2012.732140.

Vitak, Jessica, Yuting Liao, Mega Subramaniam, and Priya Kumar. "'I Knew It Was Too Good to Be True': The Challenges Economically Disadvantaged Internet Users Face in Assessing Trustworthiness, Avoiding Scams, and Developing Self-Efficacy Online." *Proceedings of the ACM Human–Computer Interaction* 2, no. CSCW (November 2018): 1–25. https://doi.org/10.1145/3274445.

Warschauer, M. *Technology and Social Inclusion*. Cambridge, MA: MIT Press, 2003.

Webster, James G., and Jacob Wakshlag. "Measuring Exposure to Television." In *Selective Exposure to Communication*, edited by Dolf Zillmann and Jennings Bryant, 35–62. Mahwah NJ: Routledge, 1985.

Weinstein, Emily. "The Social Media See-Saw: Positive and Negative Influences on Adolescents' Affective Well-Being." *New Media & Society* 20, no. 10 (2018): 3597–3623. https://doi.org/10.1177/1461444818755634.

Wetsman, Nicole. "Older Adults Struggle to Access COVID-19 Vaccine Appointment Websites." *The Verge*, January 12, 2021. https://www.theverge.com/22227531/covid-vaccine-website-appointments-accessible-seniors.

Wikipedia. "Template:COVID-19 Pandemic Data/United States Medical Cases." May 22, 2020. https://en.wikipedia.org/w/index.php?title=Template:COVID-19_pandemic_data/United_States_medical_cases&oldid=958144654.

Wohn, Donghee Yvette, and Robert LaRose. "Effects of Loneliness and Differential Usage of Facebook on College Adjustment of First-Year Students." *Computers & Education* 76 (2014): 158–167. https://doi.org/10.1016/j.compedu.2014.03.018.

Wong, Alice. "Does the Coronavirus Pandemic Make Someone Who Is Disabled like Me Expendable?" *Vox*, April 4, 2020. https://www.vox.com/first-person/2020/4/4/21204261/coronavirus-covid-19-disabled-people-disabilities-triage.

World Bank. "Individuals Using the Internet (% of Population)." 2021. https://data.worldbank.org/indicator/IT.NET.USER.ZS?view=map&year=2019.

World Health Organization. "Advice for the Public." Geneva: WHO, 2020. https://www.who.int/emergencies/diseases/novel-coronavirus-2019/advice-for-public.

World Health Organization. "1st WHO Infodemiology Conference." Geneva: WHO, 2020. https://www.who.int/teams/risk-communication/infodemic-management/1st-who-infodemiology-conference.

Yoo, Sung Woo, and Homero Gil de Zuniga. "Connecting Blog, Twitter, and Facebook Use with Gaps in Knowledge and Participation." *Communication & Society* 27, no. 4 (2014): 33–50. https://doi.org/10.15581/003.27.4.33-48.

Zillien, Nicole, and Eszter Hargittai. "Digital Distinction: Status-Specific Types of Internet Usage." *Social Science Quarterly* 90, no. 2 (2009): 274–291. https://doi.org/10.1111/j.1540-6237.2009.00617.

INDEX

Page numbers in italics refer to figures and tables.